RESOURCE WARS

RESOURCE WARS

Mapping Global Greed

BANE SINGH

Vitasta

LET KNOWLEDGE SPREAD

Published by
Renu Kaul Verma
Vitasta Publishing Pvt Ltd
2/15, Ansari Road, Daryaganj
New Delhi-110 002
info@vitastapublishing.com

ISBN 978-93-86473-95-0

First Edition 2020

MRP ₹795

Cover and layout by Somesh Kumar Mishra
Editorial Team: Manjula Lal, Saraswathi Sivakumar and Papri Sri Raman
Printed by Vikas Computer and Printers, New Delhi

Contents

To Mother Nature with due apologies

Acknowledgements

The author and the publishers acknowledge all the Open Source information used in this work.

This is a book upon natural resources of the world. Basically, it is a book on the sun, sea and the sands – ubiquitous resources that can never be fully defined. They cannot also be exclusively for the use of just one group of people. They are for the use of the entire world's people, rich or poor, black or white, privileged or under-privileged.

The definitions and descriptions of natural resources, minerals, rare earth elements, forest and water abound and nearly coincide, and therefore, cannot be copyrights of one or the other. There are umpteen standard definitions available on thousands of sites on the worldwide web; such information is for educational and mapping purposes, for the benefit of the world's student communities, inquisitive learners and workforces.

Our purpose of creating and putting in the public domain such a book is to make available to the lay reader, especially in India, one-point comprehensive reference to all the basic resources on Earth. It is a compendium of the world's physical assets and mankind's quest to acquire them, and we thank Creative Commons, Wikipedia, Wikimedia, Google Books and all other open information sources in the public domain.

Myriads of quality sources/references have been consulted while building up this mammoth work over a period of several years. Some sources have been duly credited in the various pages of this work and some are being acknowledged here also. Without consulting these highly reverential sources, this colossal work would not have been possible!

Our thanks are extended to:

Government of India and its pertinent ministry sites mnre.gov.in (Ministry of New & Renewable Energy) and State nodal agencies, ONGC (Ministry of Petroleum & Natural Gas), Ministry of Coal, Coal India Ltd, National Research Development Corporation (NRDC), Ministry of Water Resources, Ministry of Environment and Forests.

Government of the USA and its resourceful sites USGS, USEPA, USEIA, US DOE (Department of Energy), NETL, USCC.gov, EveryCRSreport.com (Congressional Research Services).

Government of Iceland and their website NEA

Government of Germany and related websites, BGR; Eurocoal, Europa.eu.

Government of China and their various websites, https://www.gov.cn/english/.

UN (United Nations) and its UNEP (United Nations Environment Programme), FAO (Food & Agriculture Organization) & UNCTAD (United Nations Conference on Trade & Development,World Investment Report), UN.org-sdg, UNESCO World Heritage Centre, World Bank.

Thanks are also extended to various media houses (print & electronic); National Geographic, BBC, Reuters, Economic Times, DW.com, the Telegraph, New York Times, CEPAL Review, Wall Street Journal, The Hindu, downtoearth, Shodhganga, guardian.com, Nuclear Engineering Handbook, globalwitness.com, MSN.com, globalnewswire.com, theglobeandmail.com, South Asia Globe, Xinhua News Agency, China Today.

Our thanks further extended to various non-profits, global associations and groups.

International Rivers (USA), IHA (International Hydropower Association), WWEA (World Wind Energy Association), IGA (International Geo-thermal Association), REN Alliance (International Renewable Alliance), EIA (Environmental Investigation Agency) London/Washington DC, WGC (World Gold Council), WGA (World Gold Association), WPIC (World Platinum Investment Council), IPA (International Platinum Group Association), ICSG (International Copper Study Group), ICA (International Copper Association), WNA (World Nuclear Association), EWEA (European Wind Energy Association), IAEA (International Atomic Energy Agency) & WCA (World Coal Association), ICID.org, NCBI-Meltdown in Tibet, hrw.org, World Nuclear Association. Thanks are also due to country grouping sites OECD (Organisation for Economic Cooperation & Development) & OPEC (Organisation of the Petroleum Exporting Countries).

Thanks further go to the organisations involved in forest-related research; WWF (World Wide Fund for nature), WRI (World Resources Institute), Sneca Creek Association, FAN

(Fundacion Amigos de la Natureleza), the Nature Conservancy & IUCN (International Union for Conservation of Nature), National Centre for Environment Information-NOAA, New Energy Institute, International Copper Study Group, mining-technology.com, National University of Singapore.

We also sincerely acknowledge various other sites for highly useful information: steelguru.com, anl.gov, Powergen International, Wave Water Works, BGR Germany, allconferences.com, leadersedge.com, Sciencedirect.com, Brookings Institute, brookings.edu, Farmlandgrab.com & CO2.earth, Foodandwaterwatch.com.

If we have left anyone out, it is not intentionally, there are countless numbers we owe our gratitude to for the mine of information available to us on the www.

REFERENCES

1 A Guide to Shale Gas, Energy Institute, England, 2015,

2 Antarctic Environmental Protocol (Madrid Protocol), 1991.

3 ARI, Advanced Resources International Report, June 2013.

4 BGR Germany Reports.

5 Bombay High Court order, September 2010 on sand ban.

6 British Petroleum Statistical Review, 2011.

7 CIA World Factbook.

8 CO2.earth.

9 Coastal Care, report on world beaches and sand mining.

10 C-WET (Centre for Wind Energy Technology) India Report on Wind Power.

11 EC (European Commission) Directives on CDW (construction & demolition waste).
12 Environmental Investigation Agency (EIA) report, 2012.
13 European Forest Institute, Finland.
14 European Photovoltaic Industry Association (EPIA) report on wind energy.
15 FAN (Fundación Amigos de la Naturaleza) report.
16 FAO Report on Coping with Water Scarcity, 2007.
17 Forest Certification standard, Sustainable Forestry Initiative (SFI).
18 Forest Trends Report, Washington DC.
19 GEA (Geothermal Energy Association) Report, 2016.
20 Global Energy Statistical Book, 2019.
21 Global Forest Resources Assessment reports, FAO, UN.
22 Guardian report through Greenpeace on illegal felling in Brazil, May 2014.
23 IAE (International Energy Agency) Report, Paris, 2013.
24 ICA (International Copper Association) report on copper.
25 ICSG (International Copper Study Group) reports, various years.
26 Indian Wind Energy Association.
27 Indus Water Treaty 1960, MEA (Ministry of external affairs) India.
28 Inter Governmental Panel on Climate Change (IPCC) Report, 2013.
29 International Atomic Energy Agency (IAEA) Report, 2017 on International Status and Prospects of Nuclear Power.
30 International Atomic Energy Agency Reports, various years.
31 International Rivers, report on China dam building.

32 IWCC (International Wrought Copper Council) report on copper use.

33 Madras High Court order, December 2010, prevention of illegal mining.

34 Mekong River Commission, Phnom Penh.

35 National Institute of Wind Energy India.

36 National Oceanic and Atmospheric Administration (NOAA).

37 Nuclear Energy Agency (NEA) reports, various years.

38 OECD (Organisation for Economic Co-operation & Development), Paris.

39 OSINFOR (Agency for the supervision of forest resources) Peru, Report on forests.

40 PriceWaterhouseCoopers reports on various emerging countries.

41 Production of Palladium, 2017, worldatlas.

42 Riding a Wave of Growth-Global Wealth 2014, BCG.

43 Rise of Renewable Energy, REI (Renewable Energy Institute), Feb 2019.

44 RRF, Resources for the Future, Washington DC Rerort on Fracking.

45 Russia TV report on Romanian Resistance against Fracking.

46 Sand & Sustainability, UNEP Report 2019.

47 Savcor Indufor Oy Report on Forests.

48 Seneca Creek Associates reports on forests.

49 Southern Africa Resource Watch.

50 The Global Wealth Report 2018-Credit Suisse.

51 The National Green Tribunal (NGT) order, August 5, 2013 banning sand mining.

52 The Supreme Court of India order, February 2012, on sand mining.
53 UN Convention on Non-navigational uses of International WCs, 1997.
54 UN law of Trans-boundary Aquifers.
55 UNCTAD-Commodities at a glance – Shale gas, 2018.
56 UNEP-IRP Global Resources Outlook, 2019.
57 USEIA (US Energy Information Administration), various years reports.
58 USEPA, US Environmental Protection Agency.
59 USGS (United States Geological Survey), various reports, Reston, US.
60 World Coal Association, various years reports.
61 World Gold Council, UK reports on gold.
62 World Nuclear Association(WNA), various years reports.
63 World Platinum Invest Council report on Platinum.
64 World Population Review, US Census Bureau.
65 World Wide Fund for Nature (WWF), various reports.
66 WRI (World Resources Institute), various reports.
67 www.gsma.com, mobile subscribers.
68 www.worldatlas.com countries with most natural resources.

Preface

Mankind has opened up all the conceivable fronts upon this planet in the pursuit of natural resources and is now raring to take a giant leap further to exploit extra-terrestrial resources! Even asteroids could be mined half a decade from now, a science fiction dream that could turn into a world-changing reality.

Humans are now seeking a new form of 'gold at the end of the rainbow' which connotes the exploitation of outer space for water and Platinum Group Metals, rare upon this Earth yet extremely important in the manufacture of electronics and other high-tech goods.

While the global population has doubled over the past five decades, the extraction of materials has tripled! The exploitation of natural resources has accelerated over the last two decades, and accounts for more than 90 per cent of our biodiversity loss and water stress—and approximately half of Climate Change impacts. Over these last 50 years, we have not seen even once a prolonged period of stabilisation or decline in global material demand.

We *homo sapiens* have now no moral qualms and compunctions even in resorting to perilous technologies for resources extraction. China is now planning to apply in hydro-fracturing (gas fracking) the same technology that was used to detonate a nuclear bomb over Hiroshima during World War II to access its massive shale gas reserves in Sichuan province. Laboratory tests have already been conducted for the same. How devastating it would be!

No place on this planet is now virgin, whether onshore or offshore, creek or river, forest or mangrove, Arctic Ocean to islands on the Pacific—everywhere avaricious eyes have pervaded and grossly violated earth's modesty. Countries with high-tech knowhow are reaping huge benefits and on the other hand poor countries often remain largely deprived. This is resulting in unequal distribution of the benefits of resources and wreaking havoc on ecosystem health and human well-being. While extraction and consumption are growing in upper-middle-income countries, high-income countries continue to outsource resource-intensive production. An average person living in a high-income country now consumes 60 per cent more than the person living in an upper-middle income country group and over 13 times the low-income groups. The twin activities of extraction and processing of natural resources are causing immense biodiversity loss, acute water stress in several geographies and elevated greenhouse gas emissions.

China's skyrocketing demand compels it to seize and take control of new territories to further exploit natural resources. In that quest, it laid claim to Japan's Senkaku Island in 2013 and keeps putting forth claims on various other islands and maritime areas in South China Sea. Today, South China Sea is the bone of contention among seven countries within the region, namely Brunei, PRC (People's Republic of China), Taiwan (Republic of China), Vietnam, Philippines, Malaysia and Indonesia. Russia unilaterally annexed Crimea in 2014 to reap the rich harvest of the region's abundant natural resources.

In the Arctic Sea, many big players have of late swung into action following the recent discovery of oil and gas which are estimated at around 13 per cent of the earth's oil and 30 per cent of the world's natural gas. Tensions are increasing worldwide, trade wars are escalating, and so are conflicts. The melting Arctic adds to the explosive cocktail—Russia, China and the US could well find themselves embroiled in another Cold War over these natural resources. The three powers have started a race to gain influence and control in the Arctic region. At stake are as much as $35 trillion worth of untapped oil and natural gas, valuable minerals including gold, silver, diamond, copper, titanium, graphite, uranium and rare earth elements that could soon be within reach as the ice recedes. With the largest number of icebreakers and $300 billion in 73 projects either completed, in motion or proposed, Russia is clearly emerging as a leader in the Arctic zone.

Africa is also blessed with many natural resources which has led to fierce competition to take control of its resources as fast as possible. China, India, France, Brazil and Russia are the main contenders in the region.

Forest resources are being plundered mindlessly. The world lost nearly 13 million hectares (m ha) of forest—an area roughly the size of England—every year between 2000 and 2010 and the world lost another 120 m ha of tree cover between 2014 and 2018, an area larger than Colombia. Illegal logging is rampant in countries like Cambodia, Bolivia, Peru, Indonesia, Ecuador, Gabon and Papua New Guinea. In Russia's Far East region, there are reports of up to 50 per cent illegal logging whereas in North-West it is 10-15 per cent. Similarly, indications are that in Brazil, illegal logging is of 20-47 per cent whereas Vietnam reports 20-40 per cent illegal logging.

Fracking has mesmerised several countries across continents. The US has been pursuing it vigorously and intrepidly and now producing around 37 billion cubic feet (bcf) per day, making it the world's largest producer of shale gas. Fracking involves injection of around 600 chemicals along with sand and water. It is said to cause many environmental threats including earthquakes, yet governments around the world—the same ones which ratified the Paris Agreement in 2015—are supporting oil and gas companies as they seek to greatly widen fracking's global footprint.

Beaches and rivers are being heavily mined for the highly coveted resource: sand. More than 70 per cent of the total world's beaches are now on the verge of complete vanishing. Rampant sand mining has become a global menace, most of it illegal, in at least five dozen countries spread over five continents. The curse is palpable from Vietnam to Jamaica in the Caribbean Sea. It has also gained media attention in several geographies that include Morocco, India, South Africa, Benin, Liberia, Grenada, Jamaica, Malaysia and Indonesia. Beach sand mining has also been documented in Mexico, Canada and Australia and is happening in numerous additional countries right now.

Water is the most important of all natural resources, the scarcity of which is being felt across the globe, particularly of potable water fit for drinking. A 2014 survey of the world's 500 largest cities reveals that one in four is in a situation of water stress. By 2040, it is predicted that 33 countries would face extremely high water stress including 15 in the Middle East, most of northern Africa, Pakistan, Turkey, Afghanistan and Spain. Countries like India, China, South Africa, USA and Australia will face high water stress. Much of the available water, both surface and ground, has been rendered contaminated due to anthropogenic activities. Irony is that many a city has been witnessing drought and flood in the same year! Mindless harnessing of water, its use and abuse, changing rainfall patterns induced by Climate Change have resulted in water scarcity in several geographies. Recently, the story of South Africa's Cape Town running out of drinking water has already hogged considerable global attention and more cities like Sao Paulo, Bangalore, Beijing, Cairo, Jakarta, Moscow, Istanbul, Mexico City, London, Tokyo and Miami are most likely to run out of drinking water unless cogent measures are taken. Water wars and struggle for water are ubiquitous; some are fought in courts, some on streets.

Domestic Extraction (DE) per capita has reached mind-boggling levels of around 100 tonne per capita in Australia! Other countries like Canada, China, Russia, USA, Brazil, Turkey, Vietnam, Indonesia and India are extracting natural resources at higher rates. These elevated levels of extraction from Earth are neither desirable nor sustainable in the long run.

Across continents, stories and narratives of rapacious exploitation of natural resources are plentiful. Lure and lust for natural resources has been mounting and escalating with no end in sight. The pages of this book are going to unravel umpteen such narratives known, unknown and lesser known that would not only startle the readers but significantly enhance their knowledge and comprehension of the most sought after and interesting topic of the current century: Natural Resources.

Introduction

Catapulting competitive drive and mounting lure for resources and wealth in the form of mineral, metal and energy coupled with augmented fears about depletion of non-renewable resources are now driving humankind to seek resources even in the outer planets and asteroids scattered in the cosmos. Newer discoveries point out enormous wealth in the NEAs (Near Earth Asteroids). Asteroids such as Eros, Aten, Phaethon, Armor and Amun are believed to contain mammoth quantities of rare Platinum Group Metals (PGMs), nickel, titanium, magnesium, cobalt, iron and many more that would provide for this planet's burgeoning needs for many thousands of years!

A stage has already been set in the formation of an offbeat company called Planetary Resources backed by bigwigs like Google executives Larry Page and Eric Schmidt, Ross Perot, James Cameron and other luminaries. It is highly likely more such companies would crop up in times to come. Human greed and need are in fact never satiated. At the same time, innovations and new avenues are never exhausted.

At the culmination of the preceding millennium, the world witnessed two great middle class expansions. The industrial revolution of the nineteenth century created a substantial Western European and American middle class and that was perhaps the first middle class expansion. Later on came the second, which sprouted after World War II, a spurt which embraced the whole of Japan. This time around, it is happening in the emerging markets (EMs) and developing countries and we are living through a third one in which the main protagonists, no doubt, are both India and China.

Globally speaking, today roughly a quarter of the world's population can be put under the middle class bracket which earns $10-$100 per day. Out of this, 60 per cent come from the developed world and 20 per cent live in the BRIC (Brazil, Russia, India and China) countries. But this scenario is going to change drastically by 2030. Just take the case of China, which has around 150 million people in the middle class. By 2030, around one billion people in China could be middle class–as much as 70 per cent of its projected population. Similarly, India's middle class, though at present much smaller at around 50 million people, or 5 per cent of the global population, will grow steadily in coming decades, reaching 475 million people by 2030.

This emerging middle class the world over has unfolded a new fascinating story on this planet by the dawn of the 21st century. The rivalry for resources control has been heating up and many new interesting stories have been shaping up, perhaps less documented so far. This new tale is about scramble for resources vying for the life-sustaining MFEW (Mineral/Metals, Food/Fibre, Energy and Water). There has been an astounding lust, mounting greed and growing compulsion to seek fresh natural resources, giving rise to the intense competition among nations in the race.

'Emerging middle class' is a critical economic and social actor because it constitutes both an engine of growth and motor for consumption, fuelling the global demand for resources. Middle class expectations in emerging and developing countries are rapidly shooting up and their mind-boggling consumption patterns have rendered this planet in 'ecological deficit' condition, a state in which ecological footprints substantially exceed biological capacity.

Today, the world's 'ecological footprint' is over 50 per cent larger than what the planet can regenerate. In other words, it now takes more than one year and six months for the Earth to regenerate what we use in a single year. Obviously, this is an upshot of indiscriminate liquidation of the planet's resources. This remains a vastly underestimated threat and poorly addressed so far!

Teeming millions have been lifted out of the poverty net and several millions added to the middle class in many developing and emerging economies. They want the same material comforts—electronic gadgets, automobile, housing, white goods, FMCGs et al—commensurate with the developed worlds of North America and Western Europe. This has put immense pressure on their governments to explore additional resources in the overseas territories of the least developed/developing countries.

China's emergence on the global scene has been quite spectacular and no chapter of this book can avoid mentioning the country. This second largest economy of the world has become a world factory for processing the raw materials of the entire planet. The roar of this Asian tiger can be heard in all parts of the globe. Its presence is felt from Australia to Angola, Mongolia to Mozambique. It is further consolidating its global position by aggressively enhancing Outward Foreign Direct Investment (OFDI). It has brought about a plethora of overseas investment bewildering the whole planet!

China has expanded its investments significantly in the past 15 years with large chunks going to countries in the developing world. Chinese firms started overseas investments as early as in the 1980s when China was undergoing rapid development. In the 1990s, China's OFDI was 1-3 per cent of GDP. China's direct investment spree started way back in the mid-2000s. From an annual average of below $3 billion before 2005, OFDI flows grew to $20 billion in 2006 and more than $50 billion by 2008. In 2010, China's annual OFDI reached $60 billion, making China one of the world's Top 10 exporters of direct investment in the post-financial crisis period. By the end of 2011, China's total global OFDI stock stood at $365 billion which moved up to $614 billion in 2013. In recent years the focus of Chinese investors has started to shift to North America and Europe also and Chinese direct investment in America and Europe has increased sharply since 2009. Chinese firms now operate in at least 35 of the 50 states in the US. Chinese investment in the US has grown from less than $1 billion annually before 2008 to $5 billion in 2010. Its overseas investments are largely driven by the prime motive of getting easy access to natural resources to quench its perennial needs and spiralling domestic demand.

Equally impressive is the story of India, which is giving tough competition to China. This notion receives further boost with an absolute majority government coming at the helm of affairs. The current dispensation in India seems determined to put prime focus on infrastructure development, FDI and international commerce and cooperation. Apart from this, there are numerous other factors which make India a frontline economic contender, including abundant natural resources, diverse climatic conditions, cost competitiveness, low labour costs, large pool of skilled manpower, strong knowledge base with significant English-speaking population, young country with a median age of 30 years by 2025, formidable labour force of nearly 530 million, high proportion of population in the working age group (15-59 years) which is likely to increase from approximately 58 per cent in 2001 to more than 64 per cent by 2021. India has the potential to undergo a paradigm shift owing to its competitive stand in the world and the Indian economy may soon be on a robust growth trajectory.

A stupendous boom in mobile telephony is evident in every nook and corner of the globe. By June 2017, there were 5 billion mobile subscribers worldwide. The small handset uses at least two dozen metals and rare earths which may include aluminium, antimony, barium, beryllium, cobalt, copper, chromium, gallium, gold, indium, lead, manganese, nickel, palladium, platinum, ruthenium, rare earths, silver, strontium, sulphur, tantalum, tin, titanium, tungsten, yttrium, zinc and zirconium. How much pressure they have been exerting on global resource exploitation is just beyond imagination!

Our planet is currently inhabited by over 7.7 billion

people. Rapid growth of the human population, especially over the last three centuries, is one of the most remarkable trends in population change ever observed. More people require more resources, such as crops, seafood, forest products, energy, metals and minerals and increasingly larger economies to support economic development and rising standards of living. Population growth and the consequent increased demand for natural resources add to the already existing tight competition.

The population story becomes more complex as there is not a simple, singular relationship between the number of people and the amount of resources consumed. Affluence, or wealth per person, and the social norms of consumption are also important factors. For example, the populations of China and India are roughly 1.4 and 1.3 billion people, respectively—about four times that of the US. However, the energy consumption per person in the US is six times larger than that of a person in China, and 15 times that of a person in India. This is because the demand for resources like energy is often greater in wealthy, developed nations like the US.

Worldwide, fossil fuels (oil, coal, and natural gas) dominate our energy consumption, accounting for 85 per cent of all energy used. There was a rapid and dramatic rise in exploitation of fossil fuels in the nineteenth century with the discovery of oil and the industrialisation of economies. It expanded rapidly in the twentieth century with increased economic development, rising populations and affluence. From 1860 to 1991, energy use per person rose more than 93-fold compared to a four-fold increase of world population, indicating that rising affluence and consumption are driving energy demand.

Land use changes are also putting enormous pressure on forest resources and biosphere. Landscapes are changing worldwide, as natural land covers like forests, grasslands, and deserts are being converted to human-dominated ecosystems, including cities, agriculture and forestry. From 2000 to 2010, approximately 13 m ha of land were converted each year to other land cover types. Developed regions like the US and Europe experienced significant losses of forest and grassland cover over the past few centuries during phases of economic growth and expansion. More recently, developing nations have experienced similar losses over the past 60 years, with significant forest losses in biologically diverse regions like Southeast Asia, South America and Western Africa.

Many countries harbour the ambition of becoming super-economic powers. All the developed and developing countries are keen on greater economic engagement with neighbouring as well as distant geographies. It is highly unlikely that the torrid race for economic supremacy would cool down in the near future. In the global race for seeking natural resources, there are established advanced economies of G-7 Group (US, Japan, Germany, UK, France, Italy and Canada), plus Australia, South Korea and Spain. Apart from this, there are seven largest emerging market economies of E-7 (China, India, Brazil, Russia, Indonesia, Mexico and Turkey). As per the latest report of Price Water House Coopers, China, US, India, Japan and Russia would be the Top 5 economic powers of the world as per projected GDP at purchasing power parity (PPP) in 2030. The next 15 slots would be occupied by Brazil, Germany, Mexico, UK, France, Indonesia, Turkey, Italy, South Korea,

Spain, Canada, Saudi Arabia, Australia, Poland and Argentina in that order. It may be noted that six Asian countries would figure among Top 20 nations by 2030. The report also reveals no African country would occupy the coveted slot of Top 20 in 2030, though Nigeria could figure in 2050. Similarly, another dark horse Vietnam can catapult itself to occupy a respectable position in the prestigious club of Top 20 by 2050. Malaysia, a potential fast-growing medium-sized economy in the Asia-Pacific region may provide a suitable launch pad for some western companies investing in the region and could become an economic power of reckoning.

Well, all these are future projections and some countries may dazzle the world in coming decades. We will continue to track future developments and perhaps get a fresh opportunity to revisit and rewrite the whole story!

Greed Unlimited

The term 'Natural Resources' confuses people. 'Natural Resources' are not like a finite number of gifts under the Christmas tree. Nature is given, but resources are created.

—Alex Tabarrok

THE GREAT Industrial Revolution in the eighteenth century initiated a dramatic shift in natural resource use which has been continuing relentlessly, unceasingly. Mankind suddenly found a hidden wealth of energy on this planet and use of fossil fuel accelerated, commencing with coal and subsequently oil and gas.

At the dawn of the twenty-first century, many nations went virtually berserk in the matter of extraction of natural resources. Only Antarctica, where mining is prohibited by the Madrid Protocol, has been left untouched as the 'Natural Reserve' of our planet. Lifestyle patterns demand production of more goods and in turn entail enhanced extraction. Increase in resource extraction is being witnessed across all major categories encompassing biomass, fossil fuels, metal ores and industrial and construction minerals. Global gas extraction, sand and gravel mining, have almost doubled in a short span of twenty-five years between 1980 and 2005, whereas nickel ore extraction has tripled during the same period. Due to overexploitation, fish catch rate has now significantly reduced.

Renowned economist Erich W Zimmermann once proclaimed in the 1930s, 'Resources are not; they become.' What he was trying to emphasise was that resources are not fixed things that merely exist, but that their meaning and value emerge as humans appraise their worth and evolve the technical and scientific knowledge to transform them into useful commodities. The importance of resources thus depends upon 'cultural appraisals', and the value of their distribution and use changes. What we presume to be a resource today may or may not remain so in the future or distant future. Technological innovations in transportation, advanced and enriched extractive technologies, as well as marketing play critical roles in changing the nature of resource use. Resources may be used, overused or abused, and the use of resources may perhaps degrade the environments in which people live. Standards of living as well as quality of life are immensely affected by the manner in which resources are utilised.

Before we launch into a serious discussion on natural resources which readers would find in ensuing chapters, it would be worthwhile and contextual to refresh our understanding about the issues involved.

A 'resource' is more or less a cultural concept. A resource is any physical material constituting part of Earth that people need and value. Natural materials palpably become resources when humans start to value them. The uses and values of resources change from geography to geography, culture to culture and of course from time to time. Resources are spatially distributed, varying in quantity and quality. Some resources are finite, while others can be replenished at varying rates. However, humans need to balance short-term rates of use against long-term availability to ensure a sustainable future.

Different categories of biotic materials such as cereals and timber and abiotic materials such as minerals, metals, fossil fuels, energy, water, land area and fertile soil together constitute natural resources. Three key basic resources—land, water, and air—are inevitably essential for survival. The characteristics and quantity of a resource are defined by whether it is a renewable, non-renewable or flow resource. Renewable resources can be replenished if their environments remain intact. Non-renewable resources can be extracted and used only once. Flow resources, such as water, wind and sunlight must be used when and where they occur because they can't be strictly classified as either renewable or non-renewable.

RENEWABLE RESOURCES

Unlike fossil fuels, renewable resources can be regenerated or replenished and they form an important aspect of sustainability. The most frequently used renewable resources world over are biomass, water, geothermal, wind and solar energy. It is pertinent to note that with the rising cost and decreasing availability of non-renewable fossil fuels, renewable resources are receiving increasing attention.

BIOMASS

Biomass comes from plants, crops, forest residues, corn kernels and stalks, energy crops, perennial grasses and fast-growing trees like poplar, to name a few. It can be used to make liquid bio-fuels that serve as alternatives to oil, or to produce heat or electricity to power our homes. Biomass power accounts for roughly half of all the renewable energy produced in the US.

Plants capture and store the sun's energy as they grow. Today's biomass energy comes from annual row crops such as corn and soybean, agricultural leftovers such as rice husk and pressed sugarcane, and wood, including whole trees, from forests—a highly problematic source of biomass. Researchers are developing ways to produce energy from special, fast-growing, and higher-yielding 'energy crops' such as willow, miscanthus and switch grass.

All this plant material can be treated in different ways to produce energy and fuel. Biomass can be burned in power plants to produce heat or electricity, with fewer harmful emissions than coal, fermented to produce fuel like ethanol for cars and trucks, digested by bacteria to create methane gas for powering turbines. Also, they can be heated under special conditions, or 'gassified', to break down into a mix of gases that can be burned for electricity or used to make a range of products from diesel to gasoline to chemicals. Apart from this, they have the innate potential to reduce greenhouse gas emissions. The agriculture and forestry industries also benefit from the demand for biomass.

GEOTHERMAL ENERGY

Geothermal energy comes from harnessing heat from the Earth. A large utility company, for example, can directly use a geothermal reservoir to drive generators and produce electricity for their municipality. In contrast, residential heat pumps use the shallow ground temperature of the Earth to heat and cool a home on a smaller scale. The shallow ground temperature remains 50-60 degrees Fahrenheit. Other applications put geothermal heat to use in commercial buildings, roads, agriculture and industrial factories. Iceland has made tremendous progress in terms of geothermal energy.

WATER

Of all the resources, water perhaps is the most sought after. Despite being the most common substance on Earth, 97 per cent is sea water, unfit for human use. Two-thirds of the fresh water is locked up in glaciers, leaving only 1 per cent for human consumption; this 1 per cent is increasingly threatened by pollution. Moreover, 97 per cent of unfrozen freshwater supply is drawn from groundwater. Globally, the major demand for water emanates from the irrigation sector (70 per cent), followed by industry (22 per cent) and domestic consumption (8 per cent only).

The world's water consumption rate is doubling every 20 years, two times the rate of population growth. With persistent regional droughts, shift of the growing population to cities, and industrial growth, it is projected that by the year 2025, water demand will exceed supply by 56 per cent.

Since 1950, thc world population has doubled and water use has tripled. As of now, only 20 per cent of the world population enjoys access to running water; over one billion people in developing countries have inadequate access to water. Geographically, there is a sharp contrast in water availability and consumption pattern. The average American individual uses 100-176 gallons of water per day whereas the average African family uses 5 gallons. At least 1 billion people walk three hours or more to obtain drinking water; individuals spend more than 200 million hours per day walking to collect water from distant, often polluted sources.

HUMAN CONSUMPTION NEEDS

Though the daily requirement of drinking water per person is 2-4 litres, it takes around 2,000-5,000 litres of water to produce one person's daily food. The UN suggests 20-50 litres of safe freshwater a day for drinking, cooking and cleaning. Water requirements of everyday eatables are abnormally high, it takes 1,000-3,000 litres of water to produce one kilogramme of rice, around 13,000-15,000 litres of water to produce one kilogramme of grain-fed beef, 35 litres of water for a cup of tea, 70 litres of water for an apple, 75 litres of water for a glass of beer, 120 litres of water for a glass of wine, 140 litres of water for a cup of coffee, 170 litres of water for a glass of orange juice, 184 litres of water for a bag of potato crisps, 200 litres of water for a glass of milk, 2,400 litres of water for a hamburger.

Rapidly growing demand for meat and milk in urban areas of developing countries will place substantial new demands on agricultural water resources, especially for growing fodder.

Water as a source of energy

Water is also the renewable energy source that produces hydroelectricity across the globe. Like wood, water has a long history as an energy source. Paddle wheels used to grind grain are an early example. In the 1880s, the Wolverine Chair Factory in Michigan made use of a water turbine and the first hydroelectric plant in the US was built on Wisconsin's Fox River to harness the power of swiftly-moving water. Hydroelectric power plants proliferated with the ability to transmit electricity over longer distances. The release, as needed, of water stored in reservoirs behind dams produces electricity by spinning turbine blades as it flows through penstocks.

SOLAR ENERGY

The sun has produced energy in the form of heat and light since the Earth formed. Solar energy systems do not produce emissions and are often not harmful to the environment. Thermal solar energy can heat water or buildings. Photovoltaic devices, or solar cells, directly convert solar energy into electricity. Individual solar cells grouped into panels range from small applications that charge calculator and watch batteries, to large systems that power residential dwellings. Photovoltaic (PV) power plants and concentrating solar power plants are the largest solar applications, covering acres.

WIND POWER

As long as the planet sustains, wind will remain an infinite, renewable resource. Wind power is clean energy because wind turbines do not produce any emissions. The classic Dutch windmill harnessed the wind's energy hundreds of years ago. Modern wind turbines with three blades dot the landscape today, turning wind into electricity. It is the fastest-growing source of new electric power.

NON-RENEWABLE RESOURCES

Non-renewable resources are resources that do not renew

themselves at a sufficient rate for sustainable economic extraction in meaningful human timeframes. An example is carbon-based, organically-derived fuel. The original organic material, with the aid of heat and pressure, becomes a fuel such as oil or gas. Fossil fuels such as coal, petroleum and natural gas and certain aquifers are all non-renewable resources.

Metal ores are other examples of non-renewable resources. The metals themselves are present in vast amounts in the earth's crust, and are continually concentrated and replenished over millions of years. However their extraction by humans only occurs where they are concentrated by natural processes such as heat, pressure, organic activity, weathering and other processes enough to become economically viable to extract. These processes normally take enormous length of time extending over tens of thousands to millions of years. As such, localised deposits of metal ores near the surface which can be extracted economically by humans are non-renewable in human timeframes, but on a world scale, metal ores as a whole are inexhaustible, because the amount vastly exceeds human demand on all timeframes.

Flow resources can be used and replenished at the same time. An example of a flow resource would be a river current. Other examples include tides, wind and solar radiation.

RARE EARTH ELEMENTS (REEs)

'Rare Earth' is often a misnomer as they are not as 'rare' as their name implies. Thulium and lutetium are the two least abundant REEs but they each have an average crustal abundance that is nearly 200 times greater than the crustal abundance of gold. However, these metals are very difficult to mine because it is unusual to find them in concentrations high enough for economical extraction. The most abundant REEs are cerium, yttrium, lanthanum and neodymium. They have average crustal abundances that are similar to commonly used industrial metals such as chromium, nickel, zinc, molybdenum, tin, tungsten and lead but they are seldom found in extractable concentrations.

REEs are a group of 17 chemical elements comprising yttrium and 15 lanthanide elements (lanthanum, cerium, praseodymium, neodymium, promethium, samarium, europium, gadolinium, terbium, dysprosium, holmium, erbium, thulium, ytterbium, and lutetium). Scandium is found in most REE deposits and is sometimes classified as an REE.

REEs are all metals and the group is often referred to as 'rare earth metals.' These metals have many similar properties that often cause them to be found together in geologic deposits. They are also referred to as 'rare earth oxides' because many of them are typically sold as oxide compounds.

GROWING DEMAND

REEs find application in almost all our new gadgets, gizmos and new technological inventions. Rare earth metals and alloys that contain them are used in many devices that people use every day such as computer memory, DVDs, iPods, iPads, rechargeable batteries, cell phones, fluorescent lighting catalytic convertors and many more. During the past twenty years, there has been an explosion in demand for many items that require rare earth metals. Two decades back, there were very few cellphones in use, but the number has risen to over 7 billion today. The use of REEs in computers has grown phenomenally. Many rechargeable batteries are now made with rare earth compounds. Demand for the batteries is being driven by demand for portable electronic devices such as cellphones, readers, portable computers and cameras. Several pounds of rare earth compounds are in batteries that power every electric vehicle and hybrid-electric vehicle. As concerns for energy independence, Climate Change and other issues drive the sale of electric and hybrid vehicles, the demand for batteries made with rare earth compounds would mount even faster. Rare earths are used as catalysts, phosphors, and polishing compounds. These are used for air pollution control, illuminated screens on electronic devices, and the polishing of optical-quality glass. All of these products are expected to experience rising demand.

Other substances can be substituted for REEs in their most important uses; however, these substitutes are usually less effective and costly. From the 1950s until the early 2000s, cerium oxide was a popular lapidary polish. It was inexpensive and very effective. The recent price increases have almost eliminated the use of cerium oxide in rock tumbling and the lapidary arts. Other types of polish, such as aluminium and titanium oxide, are now used in its place.

USES IN DEFENCE

REEs also play an essential role in the defence sector. Military uses encompass night-vision goggles, precision-guided weapons, communications equipment, GPS equipment, batteries and other defence electronics. Rare earth metals are key ingredients for making the very hard alloys used to manufactured vehicles and projectiles that shatter upon impact in thousands of sharp fragments.

The table given below reflects other defence uses of REE, bringing forth their importance:

Other Defence Uses of Rare Earth Elements	
Lanthanum	night-vision goggles
Neodymium	laser range-finders, guidance systems, communications
Europium	fluorescents and phosphors in lamps and monitors
Erbium	amplifiers in fiber-optic data transmission
Samarium	permanent magnets that are stable at high temperatures precision-guided weapons 'White Noise' production in stealth technology

Source: All Infographs in this book have been made by the author through Chart Wizard after culling/filtering/converting data into desired units. No graph has been directly copied from any single source; in most cases the sources have been provided. This has been done for reader facilitation.

GEOGRAPHY-SPECIFIC RESOURCES

Lithium Triangle: Argentina-Chile-Bolivia

The majority of global lithium production comes from continental brines (also known as salt lakes, salt flats or Salars) in what is known as the Lithium Triangle—a region that includes parts of Argentina, Chile and Bolivia.

The best example is the 3,000-sqkm Salar de Atacama in Chile, which contains the highest lithium concentration and estimated lithium resources of 6.3 million tonnes (mt). Two of the world's leading lithium producers, Sociedad Quimica y Minera and Rockwood Holdings, operate on the Salar de Atacama. The Salar de Atacama produces more than half of the world's total lithium.

The Salar del Hombre Muerto, or Dead Man's Salt Flat, in Argentina is another important source of lithium at around 4,000 metres above sea level. Argentina, Chile and Bolivia together hold the planet's largest reserves of lithium, the world's lightest metal and a key component in batteries used to power a range of technologies from cellphones to laptops to electric cars. Industrial production from countries in this so-called Lithium Triangle is already high. Chile is the world's leading source of the metal, providing around 40 per cent of global supply, and Argentina is also a significant producer. Bolivia holds an estimated 50 per cent of the world's lithium reserves.

AUSTRALIAN OUTBACK

The Outback generally refers to the vast, remote and arid area of Australia. This region is comparatively more remote than those areas named the 'Bush' which normally refers to any lands outside the main urban settlements.

The main economic activity in this vast and thinly populated area is mining. The Outback is immensely rich in iron, aluminium, manganese and uranium ores, and also contains massive deposits of gold, nickel, iron, lead and zinc ores. Major mines and mining areas include opal at Coober Pedy, Lightning Ridge and White Cliffs, metals at Broken Hill, Tennant Creek, Olympic Dam and the remote Challenger Mine. Oil and gas are extracted in the Cooper Basin around Moomba.

The Pilbara region's economy activity is dominated by petroleum and mining industries. Most of Australia's iron ore is also mined in the Pilbara and it also has one of the world's major manganese mines.

The Argyle diamond mine in the Kimberley is the world's biggest producer of natural diamonds and contributes nearly one-third of the world's natural supply.

ALBERTA (CANADA) TAR SANDS/OIL SANDS

Largely located in Alberta (Canada), these tar sand deposits are distributed over an area of 140,200 sq km. Alberta ranks third after Saudi Arabia and Venezuela in terms of proven global crude oil reserves. In 2017, Alberta's total proven oil reserves were 165.4 billion barrels (bbl).

Almost 99 per cent of Alberta's proven oil reserves are found in sand; the rest comes from conventional crude oil. Alberta accounts for an overwhelming majority (about 98 per cent) of Canada's oil reserves.

The oil sand area encompasses the Athabasca, Cold Lake and Peace River areas in northern Alberta. Together, they contain an estimated 1.84 trillion barrels of crude bitumen. About 9 per cent of this volume is recoverable using extant technology and is considered to be a proven reserve.

Canada's tar sands are the biggest energy project in the world, currently producing 1.9 million barrels of oil a day. Already, millions of barrels of tar sand oil have been extracted from under the Canadian wilderness, producing 3-4 times more greenhouse gas emissions than conventional oil extraction and using enough natural gas every day to heat three million Canadian homes. The project is also causing massive deforestation. Though tar sand development has been labelled 'the most destructive project on Earth', yet the Canadian government counters all the allegations.

AMERICAN and CHINESE SHALE FIELDS

Shale gas is a natural gas from shale formations. The shale acts as the source as well as the reservoir for the natural gas. Older shale gas wells were vertical while more recent wells are primarily horizontal and need artificial stimulation, like hydraulic fracturing. Not all shale formations yield gas.

Shale gas is one of the fastest growing forms of natural gas. Along with other non-conventional forms of natural gas, such as tight gas and coal bed methane (CBM), it will make a major contribution to future North American gas production. Unconventional gas production is forecast to increase from 42 per cent of total US gas production in 2007 to 64 per cent in 2020.

American shale gas reserves are spread across the geographies of Antrim (Michigan), Barnett (Texas), Duvernay (Alberta), Eagle Ford (South Texas), Haynesville (Northern Louisiana/ Eastern Texas), Horn River (British Columbia), Marcellus (Appalachian basin North America), Caney (Oklahoma), New Albany (Illinois) and Utica (New York/Ohio).

Areas in Appalachia, Northern Louisiana, British Columbia and South Texas indicate the widespread potential of shale gas resources across North America. The recent advancements in drilling and well completion technologies, including hydraulic fracturing, has opened up plays in a number of different basins of the USA and elsewhere in the world that were not previously reckoned to have economic potential.

Chinese Shale Gas Potential

As per the US Energy Information Administration (USEIA), although production in China was not quite impressive until 2013, the volume of recoverable shale gas in China has been estimated to the tune of 1,115 trillion cubic feet (TCF) the largest of any country in the world. As of 2019, China is one of only four countries apart from the US, Canada and Argentina to produce shale gas in commercial quantities. Algeria and Mexico would start production in 2020 and 2040 respectively.

The majority of Chinese shale reserves are concentrated in

three basins of Sichuan, Tarim and Yangtze accounting for 89 per cent of the estimated national reserves.

One of the major challenges for shale gas exploration in China has been the availability of water which is the key element for hydraulic fracking. Sinopec (China Petroleum and Chemical Corp) and Royal Dutch Shell are leading the pack among various companies to meet extraction targets.

ARCTIC OIL & GAS RESERVES

Arctic resources received much attention in 2013 when Greenpeace ship Arctic Sunrise left the Norwegian port (11 August 2013) to begin a month-long expedition in the Arctic to protest against oil exploration and digging. Twenty-six activists were arrested by the Russian government and released in December 2013.

The region above the Arctic Circle accounts for only about 6 per cent of the Earth's surface area, but it could account for nearly one-fifth of the world's undiscovered yet recoverable oil and natural gas resources. The existence of hydrocarbon resources in the Arctic has been known for long. However, full-scale resource development and navigational activities began only in 2013 when the Russian firm Gazprom started oil production in Prirazlomnoye field in Arctic offshore oilfield. Since then activity has been stepped up.

Nearly 33 per cent of the Arctic is covered by land; another third consists of the offshore continental shelf and the remaining third comprises ocean waters, typically deeper than 500 metres. Much of the Arctic water is ice-capped for most of the year. However, the polar ice cap has been noticeably retracting in recent years, quite probably as a consequence of global warming.

The Arctic region contains parts of eight countries—Canada, Denmark (Greenland), Finland, Iceland, Norway, Russia, Sweden and the United States. Finland and Sweden do not border on the Arctic Ocean and are the only Arctic countries without jurisdictional claims in the Arctic Ocean and contiguous seas.

Large Arctic oil and natural gas discoveries commenced in Russia in 1962 with the discovery of the Tazovskoye field, followed by the US Alaskan Prudhoe Bay field in 1967. Approximately 61 large oil and natural gas fields have been discovered so far within the Arctic Circle. Of these 61 fields, 43 are in Russia, 11 in Canada, 6 in Alaska and 1 in Norway.

The United States Geological Survey (USGS) disseminated the first-ever wide-ranging assessment of Arctic oil gas resources in 2008 reflecting the region's undiscovered and technically recoverable conventional oil and natural gas resources. Of the 33 Arctic sedimentary 'provinces' that the USGS evaluated, 25 were found to have a greater than 10 per cent probability of having oil or gas deposits larger than 50 million barrels of oil equivalent (boe). The USGS assessment was that approximately 90 billion barrels (bbl) of oil, 1,669 tcf gas, and 44 bbl of natural gas liquids (NGLs) may remain to be found in the Arctic. Of the total 412 boe, approximately 84 per cent is expected to be found offshore, and about two-thirds (67 per cent) of the total is natural gas.

INDIA'S KRISHNA-GODAVARI BASIN

Deltaic plains formed by two large east coast rivers, Krishna and Godavari, in the state of Andhra Pradesh and the adjoining areas of Bay of Bengal in which these rivers discharge their water is known as Krishna Godavari (KG) Basin. The major geomorphologic units of the KG Basin are upland plains, coastal plains, flood and delta plains. The Basin is an established hydrocarbon area with a resource base of 1,130 mt, of which 555 mt are assessed for the offshore region. Several oil and gas fields are located in both inland and offshore parts of the basin.

Reliance Industries Ltd (RIL) is exploring oil and gas here in partnership with British Petroleum (BP). Inland, it covers an area of approximately 15,000 sq km and the offshore part covers an area of 25,000 sq km. The basin contains about 5 km thick sediment with several cycles of deposition, ranging in age.

In view of the fact that hydrocarbon indications are observed in well KB-4B-1, drilled in the north-western part of offshore basin, and also, in well KG-1B-1, indication of gas with higher hydrocarbon and oil stains in ditch samples collected from Late Paleozoic sediments, imparts the older sequence a fair degree of importance. These older sediments can also be expected to be present up to Krishna Island area around the coastal part. Gas fields like Mandapeta and Endamuru and indications of hydrocarbons in offshore areas have great potential.

Hydrocarbons are present in several fields e.g., Kaikalur, Bantumilli, Lingala, Narsapur and Razole Chintalapalli both in East as well as West Godavari Sub-Basins. The cretaceous sequence in offshore wells, like well KB-1B, has also indicated presence of hydrocarbons during drilling. Suitable source and reservoir facies (composition and fossil content) are also reported in this well (See Figure 10). Some twenty commercial accumulations have been discovered so far.

A number of gas fields are producing from Paleocene reservoirs, particularly in East Godavari Sub-Basin. The

Tatipaka, Pasarlapudi, Kadali and Manepalli are the fields located on land, while GS-8 is offshore. The hydrocarbon generation centres in Paleocene reservoirs indicate fair to rich organic content on the basinal side. The indications of gas and its pressure in this sequence justify good potential for Paleocene in the basin. Ten pools of hydrocarbon have already been discovered in this age group.

The Eocene accumulation of gas is observed in Elamanchili, Tatipaka, and Pasarlapudi. Mori oilfields including GS-38 in offshore area indicate good hydrocarbon potential. The Mio-Pliocene sequence in offshore part is promising. The commercial hydrocarbon accumulation in Ravva field is well known. The prospects GS-38, G-1 and G-2 are also hydrocarbon bearing in Mio-Pliocene strata. As many as fourteen commercial finds have come from this sequence. The site Dhirubhai-6 (D6) is where Reliance Industries discovered the biggest gas reserves in India. In government records, the 7,645 sq km block is known as KG-DWN-98/1. The KG Basin is considered to be the largest natural gas basin in India.

CENTRAL ASIAN OIL AND GAS RESERVES

Central Asian countries of Kazakhstan, Tajikistan, Kyrgyzstan, Uzbekistan and Turkmenistan, once part of the Soviet Union, have been independent since 1991. All these five states have considerable oil and gas reserves and the first four have coal reserves of significant reckoning as well. As of now, the full potential has not been realised owing to lack of technological advancement or support, absence of strategic partner and fund shortage. A brief elucidation of all these reserves is enumerated below:

CENTRAL ASIAN OIL RESERVES

Country	Proven Oil Reserves(barrels)	World Ranking
Kazakhstan	30,000,000,000	11
Turkmenistan	600,000,000	44
Uzbekistan	594,000,000	47
Kyrgyzstan	40,000,000	81
Tajikistan	12,000,000	88

*Estimates as of 2011
**Source: CIA World Factbook

Of the five countries in the Central Asian region, Kazakhstan commands the highest proven oil reserves at 30 bbl, putting it at eleventh rank in the world. It also has three huge oil fields that have 500 or more million barrels of proven recoverable oil, from which it extracts the bulk of its oil.

The next two Central Asian countries with the largest oil reserves belong to Turkmenistan and Uzbekistan that are currently producing 198,150 barrels per day (bpd) and 70,880 bpd respectively.

CENTRAL ASIAN NATURAL GAS RESERVES

Country	Gas Reserves in billion cum	World Ranking
Turkmenistan	7504	6
Kazakhstan	2407	15
Uzbekistan	1841	20
Kyrgyzstan	5.663	90
Tajikistan	5.663	92

*Estimates as of 2010
**Source CIA World Factbook

When it comes to natural gas, Turkmenistan leads the pack among its Central Asian neighbours at 7,504 billion cubic metres (bcum), making it the sixth most natural gas rich nation in the world. Turkmenistan currently produces 38.1 bcum of gas a year. However, it is ranked at only 24 among all natural gas producers, illustrating its potential in the coming years. Turkmenistan's future success in this arena will largely depend on its ability to attract foreign investment. Kazakhstan and Uzbekistan come in second and third place for natural gas reserves at 2,407 bcum and 1,841 bcum. Kazakhstan's natural gas production is behind in exploration, production and general infrastructure when compared to its oil industry. Recently Kazakhstan has become a net exporter of natural gas. Uzbekistan, on the other hand, has been developing its natural gas production for years. It is currently ahead in natural gas production in Central Asia largely due to foreign debt incurred by the government to support the incipient energy industry at the time. This has allowed the country to now rank as the fourteenth largest gas producing country in the world. Moreover, Uzbekistan's natural gas is used for both domestic consumption and foreign export. Uzbekistan currently exports to only Kazakhstan, Kyrgyzstan, Tajikistan and Ukraine due to its landlocked geographical location.

CENTRAL ASIAN COAL RESERVES

Country	Estimation Year	Proven Coal Reserves in tonnes
Kazakhstan	2009	31300 000 000
Kyrgyzstan	2003	2500 000 000
Uzbekistan	2011	1900 000 000
Tajikistan	2003	800 000 000

Source: Infographics created by author for reader facilitation.

World coal demand is likely to remain unabated till 2050 and this puts Kazakhstan in a coveted position keeping in view the mammoth coal reserves. Kazakhstan takes the lead In terms of coal, with proven reserves of 31.3 billion tonnes (bt), placing it on the top slot in Central Asia and eighth in the world. Kazakh coal reserves would last at least for the next 100 years. At the moment, the biggest importers of Kazakh coal are Russia and Ukraine. However, growing energy shortage in the world will make Kazakh coal enticing for China, Tajikistan, Kyrgyzstan, and Eastern Europe in the near future.

Next comes Kyrgyzstan with 2.5 bt coal reserves. Kyrgyzstan has a long history of coal production that started way back in the 1900s. During the Soviet Union era, state-owned coal mining companies were started to feed the Soviet economy. However, since gaining independence in 1991, Kyrgyzstan has been unable to operate many of these mining facilities due to lack of funding, infrastructure and transportation.

Uzbekistan is in the third slot among Central Asian coal reserves with 1.9 bt. The country has seen a decline in coal production—in 2009, the country produced 3.6 mt, around 25,000 tonnes less than in 2008. It was 3.40 mt in 2017 and went down to 3.2 mt in 2018. The Uzbek government has shown abiding interest in upgrading technology and improving extraction efficiency.

The world came to know of the Mongolian Tavan and Oyu Tologi Coal and Metal Reserves in 2001 when Canadian mining giant Ivanhoe Mines made its first discovery in Mongolia's Gobi desert which resulted in one of the largest finds in mining history with huge deposits of iron, gold and silver. In Oyu Tolgoi, over 81 billion pounds of copper and 46 million ounces of gold are expected to ensure production for well over 50 years. The $10 billion Oyu Tolgoi copper-gold deposit is 66 per cent owned by Ivanhoe and Rio Tinto, with the remaining portion falling to the Mongolian government. These are largely untapped resources.

Tavan Tolgoi (Five Hills) is an open coal pit expected to contain 6-7 bt of coal containing both coking and thermal coal, one quarter of the total being high quality coking coal. It is divided into six sections: Tsankhi, Ukhaa Khudag, Bor Tolgoi, Borteeg, and Southwest and Eastern coalfields. The Tsankhi section is the largest part, and is divided into East and West Tsankhi—these have been in focus recently.

All of Tavan Tolgoi is owned by Erdenes MGL (a state-owned company) except for Ukhaa Khudag section which is mined by the Mongolian Mining Corporation. Erdenes Tavan Tolgoi LLC (Erdenes TT), a subsidiary of Erdenes MGL, is managing the development of East Tsankhi. The company is due to be listed on the Hong Kong, London and Ulaan Baator stock markets sometime in the next couple of years. Mining of West Tsankhi is planned to be contracted to a consortium of international mining companies.

There are numerous other geography-specific resources: known, lesser known, tapped, partially tapped or untapped. It is not feasible to describe all of them in detail. Yet mention may be made of some of the important resources, namely Komoto Mines of DRC (Democratic Republic of Congo, Africa) containing copper and cobalt, oilfields in ultra-deepwater off the coast of Brazil, gasfields of Turkmenistan, shale gas fields of Patagonia and Neuquen (Argentina), uranium and copper mines Olympic dam (Australia), Siberian Yamal and Kara Sea oil and gas reserves

(Russia) and Kashgan oilfields of the Caspian Sea.

It becomes clear that the globe is replete with enormous resources of varying categories yet we are not extracting them in a judicious way. Mankind's rapidly growing consumption of fertile land, water, energy and materials has inflicted severe damage on Earth's ecosystem, some of them now irreversible. Our climate is changing; fresh water reserves, fish stock and forest cover are shrinking, fecundity of soil is diminishing and more and more species are becoming extinct. In order to continue to thrive on this planet, our lifestyles now ought to become more sustainable, so that we can preserve our natural resource base and the fragile ecosystems on our planet.

Territorial Ambitions, Acquisitions and Annexations

ALBATROSSES SOAR high in the azure sky, making weird sounds hither and thither in this completely deserted and barren island in the unfathomable depths of the great vast Pacific Sea. These bizarre birds seem to be reincarnations of the Lost Mariners' soul. The isolated realm becomes more eerie when dusk descends and the veil between darkness and blue-black waters becomes invisible. Surging waves desperately endeavour to kiss the terrestrial vegetation of the land, yet eternally fail. Intermingling sounds of waves, stray birds and gently blowing winds are the prelude to signs of life...

This is Senkaku in Diaoyu Islands where China unilaterally proclaimed ADIZ (Air Defence Identification Zone), assuming full sovereignty over the area on 23 November 2013. The mad race for control over newer and more resources is not sparing even far-off virgin islands. The global lust for resources is heating up, with China in the forefront followed by the US, Russia and other military-industry syndicates.

Today, China has territorial claims on almost all of its neighbours. The US often attempts to play the role of a mediator in resolving disputes in the region. But it seems that China does not care for this gratuitous service. Most of China's territorial claims betray greed for resources and maritime control. Chinese leader Mao Zedong once propounded a global goal: 'We must conquer the globe where we will create a powerful state.' China as of now has been pursing and nurturing territorial ambitions and designs, as outlined by Zedong, more aggressively and rather actively promoting the same.

Significant reserves of natural gas and oil were found on the islands and water surrounding Senkaku. For the growing industry of China and stagnant Japan, the struggle for the archipelago is serious—not to mention the fish found there in large quantities. To date, the only agreement the parties have reached in the negotiations is on the joint development of oilfields. In addition, if the Japanese behave more or less decently, the Chinese are regularly caught for illegal fishing in the area. Any resolution of a territorial dispute will set a precedent. If China's claim to even one territory from the list of the 'lost' is satisfied, the Chinese machine would discover a fresh one.

In Southeast Asia, China claims sovereignty over virtually all the waters of South China Sea and has grown increasingly aggressive in asserting claims over islands and underwater resources. In January 2014, the southern Chinese province of Hainan enacted regulations requiring all non-Chinese fishing vessels operating in the claimed waters to first obtain permission from Beijing. In 2012, China dislodged the Philippines from the Scarborough Shoal, a reef contested by the two countries. In recent years, Chinese has repeatedly detained Vietnamese fishermen and cut the cables of that country's vessels operating near island groups disputed between Beijing and Hanoi. As a result of these incessant moves, Asian countries such as Vietnam, the Philippines, Cambodia and India are now actively forging closer ties with the United States and Japan.

China seems exceptionally ready to use mild force for the sake of expansion. In the last few years, China has not only developed the industry, but also equipped its area of interest with sophisticated military equipment. China has deployed over three dozen new diesel and nuclear submarines in the region, purchased four destroyers of class 'Modern' from Russia and built another dozen on its own, and has launched a network of ground-based ballistic missiles to destroy naval targets.

SOUTH CHINA SEA

In the global race for natural resources, perhaps the South China Sea waters always remain turbulent, bringing a new controversy with every high tide. Why is it a hotbed of conflict? Why are clashes reported rather frequently? Why are all the adjoining nations in the fray? The answer is simple: the entire area is potentially rich in oil and natural gas deposits and obviously China wants a lion's share. The Chinese Mining Ministry claims there is 17.7 bt of crude oil. The US Geological Survey estimates the region's discovered and undiscovered oil reserves at 28 bbl; natural gas resource estimations range from 25.5 trillion cubic metres (tcum) to 56.6 tcum.

Territorial disputes in the South China Sea involve both

land (island) and maritime disputes among seven countries within the region namely Brunei, PRC (People's Republic of China), Taiwan (Republic of China), Vietnam, Philippines, Malaysia and Indonesia.

Disputes include the maritime boundary in the Gulf of Tonkin as well as maritime boundaries off the coasts of Vietnam, Malaysia, Brunei and the Philippines. There is a further dispute in the waters near the Indonesian Natuna Islands. Additionally, there are disputes among the various island chains of the South China Sea basin, including the Spratly Islands and the Paracel Islands. The interests of different nations include acquiring fishing areas around these two archipelagos, the potential exploitation of suspected crude oil and natural gas under the waters of various parts of the South China Sea, and the strategic control for navigation of ships of important shipping lanes.

The South China Sea is dubbed by China as the 'Second Persian Sea'. State-owned China Offshore Exploration Corporation plans to spend $30 billion in the next two decades to harness oil in the region, with the estimated production of 25 mt of crude oil and natural gas per annum at a depth of 2,000 metres.

The Nine-dotted Line: The Benchmark of Chinese Claims

The nine-dotted line was to begin with an 'eleven-dotted-line,' as indicated by the then Kuomintang government of the Republic of China in 1947 for its claims to the South China Sea. After the Communist Party of China took over mainland China and formed the People's Republic of China in 1949, the line was reviewed and changed to nine as endorsed by Zhou Enlai. The legacy of the nine-dotted line is largely held by many Chinese government officials as well as by the Chinese military, as providing historical support for their claims to the South China Sea.

Back in the 1970s, Malaysia, the Philippines, and other countries began referring to the Spratly Islands as included in their own territory. On 11 June 1978, President Ferdinand Marcos of the Philippines issued a presidential decree proclaiming the Spratly Islands as Philippine territory.

The bountiful fishing opportunities within the region are another motivation for the struggle. In 1988, the South China Sea is believed to have accounted for 8 per cent of world fishing catches, a figure that has grown since then. Many confrontations have been reported in the Philippines with foreign fishing vessels including China in the Philippine Exclusive Economic Zone. China believes that the value in fishing and oil from the sea has risen to a great high.

Heightened Maritime Activities

The South China Sea records heavy maritime traffic. The area is one of the busiest shipping routes in the world. In the 1980s, around 270 merchant ships used the route each day. As of now, more than half the tonnage of oil shipped by sea routes passes through it, a figure rising consistently with China's burgeoning oil demand. This maritime traffic is believed to be three times greater than that passing through the Suez Canal and five times more than the Panama Canal. Maritime activities are further supposed to climb in the years to come.

Current Scenario in the South China Sea

As of 2012, all of the Paracel Islands are under Chinese control and Vietnam, the Philippines, Brunei, Malaysia and other countries vociferously assert claims within the Chinese nine-dotted line. The United Nations Convention on the Law of the Sea (1994) precipitated more intense territorial disputes between the parties in the fray. Most probably, China now controls eight of the Spratly Islands and another eight are controlled by the Philippines. Vietnamese troops control 29 Spratly Islands and the rest are held by Malaysia (five), Brunei (two) and Taiwan (one).

China adheres to the policy of incremental increase by acquiring small territories so as not to invite full-fledged global attention. It recently planted an oil rig in Vietnam's EEZ (Exclusive Economic Zone) in the disputed South China Sea. China even sent military vessels to defend the oil rig, resulting in a face-off with Vietnam.

Similarly, it violated Philippines' sovereignty on Scarborough Shoal. This shoal off the coast of Zambales lies within the Philippines EEZ but has been exclusively under the control of the Chinese since the 2012 standoff. The Philippines approached the International Tribunal on the Law of the Sea (ITLOS) raising concerns over eight maritime features that are currently under the control of China. The first four are low tide elevations that are completely submerged: Mischief Reef, which is 130 nautical miles (nm) from Palawan, Kennan Reef (180 nm), Gaven Reef (205 nm) and Subi Reef (230 nm). These are all below water at high tide. They're not entitled to anything. No territorial sea, no EEZ. But all of these four are physically occupied by China, the Philippines contends, as China has constructed concrete structures on the reefs including helicopter landing pads over the years since 1995. However, that does not give them entitlement to the sea around it.

The other four are rocks or reefs that are, at most, entitled to only 12 nautical miles. These are Scarborough Shoal (120 nm from Luzon), Johnson Reef (180 nm from Palawan), Cuarteron Reef (240 nm from Palawan) and Fiery Cross Reef (255 nm from Palawan). The Philippines says China's nine-dash demarcation line is invalid and violates United Nations Convention of the Law of the Sea (UNCLOS). The four reefs that are below water all the time are entitled to no rights and the four are rocks which would only have 12 nautical miles.

CRIMEA

Russia's unquenched thirst is well known and the March 2014 annexation of Crimea has amply brought this greed to the fore. Jubilant and upbeat, Russia may perhaps crave for more territories like Transnistria, Abkhazia or South Ossetia, eastern Ukraine and ultimately Alaska! Russia's heavy engagement in the Arctic Circle is already evident where oil and gas resources are being exploited to the hilt.

The Crimean takeover brought valuable properties and vast natural resources to Russia. Following the annexation, Russia has formally scuttled the 2010 Kharkiv Agreements under which it provided a $100 discount per 1,000 cum for Ukraine's imports of Russian gas as rent for the Sevastopol naval base, home to Russia's fleet in Black Sea. However, Ukraine still regards the agreement as valid and is determined to take Russia to court for the outright violation.

One could very well call Crimea a natural museum where a variety of picturesque landscapes create such a unique combination. In many respects, it appears due to peculiarities of the peninsula's geography, geology, topography and climate.

Almost the entire natural diversity of the planet—mountains and plains, seas, rivers and lakes, forests, grasslands, subtropical and black soil, mountain and forest complexes and alkaline desert—are present in miniature in Crimea. Crimean vegetation is equally luxuriant. Diverse plant communities from mountain forests to mountain meadow and lowland steppe alternate in a relatively small area.

The entire area of Crimea comprises 78 per cent plains and 22 per cent mountains. Most of the rivers of Crimea are small rivers. The total length of the river network of Crimea is around 6,000 km. Major river flow is regulated by reservoirs and judiciously put to use for water supply and irrigation. Moreover, it is endowed with three artesian basins: North Sivash, Belogorsk, Alma. They are used for drinking water supply. In some areas, there are mineral and thermal waters used for recreation and balneal purposes.

It is home to exotic trees and shrubs. Wild rare plants constitute more than 65 per cent of the flora of the entire European part of the Commonwealth countries. Alongside, about 1,000 plant species from different continents are cultivated here. There are about 3,000 species of wild plants on the peninsula. Forest cover constitutes about 12 per cent of the geographical area. Forest plant communities are represented mainly by beech, oak, hornbeam forests and Crimean pine woods.

The fauna of the peninsula is also rich and diverse. The forests of the mountainous Crimea are inhabited by Crimean red deer, roe deer, fallow deer, wild boar, foxes, badgers and various birds, including woodcock, black and griffon vultures.

Seaside parks are amazingly beautiful! Ten of them, including Alupka, Gurzuf, Livadia, Massandra, Foros parks, are monuments to landscape art and culture.

The Nikita Botanical Garden is a real plant treasury of the peninsula. The garden over 40 hectares (ha) has over 15,000 species, forms and varieties of plants.

The area of nature reserve fund constitutes about 5 per cent of Crimea. There are 149 objects of natural reserve fund totalling more than 129,000 ha. As many as 346 deposits of fossil minerals of national and local value were explored in Crimea, 178 being currently under exploitation. As many as 159 deposits predominate in mineral-building raw materials. The second place is occupied by 87 fresh and mineral groundwater deposits. This is followed by 50 hydrocarbon deposits, 26 therapeutic mud fields, 15 deposits of raw material for the metal processing industry, 2 deposits of clay and 1 field of thermal water.

Abundant Energy Resources

Russia's natural resources ministry estimates Crimea's energy reserves at around 44 mt of oil and 165 bcum of gas; analysts estimate the value of these reserves at somewhere between $800 million and $1.2 billion. However, Crimea's natural wealth could in fact be significantly higher, with an estimated 2.3 tcum of gas in unexplored offshore fields.

Using similar methodology, such reserves may be worth around $7 billion; the value could double if the reserves are proven.

Crimea's Other Assets

No figure has yet been arrived at on the value of other properties except military equipment and natural resources, but it is likely to run into billions of dollars. Estimated value of Ukrainian military equipment left in Crimea may be around $1.7 billion. Crimea has nationalised other state property in Ukraine such as the buildings of government ministries and agencies, educational institutes and subsidiaries of state companies. The nationalised properties include 131 agro-industrial enterprises, including renowned vineyards, 130 objects of tourist infrastructure, as well as seven ports. Roughly speaking, there could be over 57,000 objects of state property in Crimea.

ARCTIC OCEAN

Many big players have of late swung into action following recent discoveries of oil and gas which are estimated around 13 per cent of Earth's oil and 30 per cent of the world's natural gas. Countries eyeing the North Pole also see it as big opportunity.

Ice has been gradually melting here as a direct indirect consequence of global warming. The game has started. Russia has dispatched a new fleet of submarines into the Arctic Ocean to gather geological samples. It already planted a titanium flag in 2007 to metaphorically support the claim that much of the sea floor as far as the North Pole itself is an extension of Siberia's continental shelf. That would expand Russia's borders to cover significant quantities of oil and natural gas.

In September 2013, Russia announced it was sending its warship to the Arctic. Russia has been reinstating its military base in the Novosibirsk Archipelago off the northern Siberian coast since September 2013.

India too has joined the hunt. ONGC Videsh entered into an agreement with Russian oil major Rosneft in May 2014. The two companies have signed a MoU by which they would form a consortium and further include other partners to explore hydrocarbon in Arctic offshore. In January 2020, India confirmed its stake in Roseneft's new Arctic oil project.

China, despite being too far from the Arctic Circle to claim territory, has expressed a sudden desire to join the once-obscure Arctic Council. The organisation—whose members include Canada, Denmark, Finland, Iceland, Norway, Russia, Sweden and the United States and representatives of indigenous Arctic peoples—was founded in 1996 to promote environmental protection and maritime safety of the Arctic. With so much natural wealth at stake, the Council's policies are increasingly critical.

The US and its northern neighbours are at odds over the Northwest Passage, a route along Arctic North America that became devoid of ice along its entire stretch for the first time in August 2007. The passage is at least 6,000 km shorter than the usual sea route through the Panama Canal, and cuts travel time by two weeks, a boon in disguise to major exporters like China. While Canada insists that part of the passage falls within its borders, the United States and Europe counter that it's an international waterway. On the other side of the Arctic Ocean, Russia claims ownership of the Northeast Passage which, since 2008, has been consistently navigable during the summer and early fall, allowing ships to take a shortcut between Asia and Europe. As of now, harsh climate of the polar region pose a great deterrent to smooth exploitation of the concealed resources but human endeavours have been continuing unflinchingly.

SOUTH KOREA

Farmland is rapidly shrinking in South Korea. Industrial development and population pressure has pushed Korea's food needs far beyond the capacity of its own land. In response, South Korea has become one of the most aggressive countries in overseas farmland acquisitions since the 2008 food price crisis along with a few other countries such as China, Japan, United Arab Emirates (UAE) and Saudi Arabia.

In order to secure an assured supply of food, the South Korean government and its multi-national corporations (MNCs) have bought millions of hectares of agricultural land in under-developed countries. Many of these acquisitions took place in South East Asia and Africa. The details of the land purchases are difficult to access, but many deals have been struck in countries with questionable human rights records such as Sudan. Most of South Korea's largest food corporations are involved in these negotiations, often with the active support of their government. In some cases, state-owned corporations such as the Korea Rural Development Corporation directly own and operate these farms. The South Korean multinational Daewoo Logistics secured 1.3 m ha of farmland in Madagascar in 2008 to grow maize and crops for bio-fuels. The deal led to widespread unrest in Madagascar and was a significant contributor to the fall of President Marc Ravalomanana. Successive governments quickly reversed this deal.

In 2009, Hyundai Heavy Industries acquired a majority stake in a company cultivating 10,000 ha of farmland in the Russian Far East and a wealthy South Korean provincial government secured 95,000 ha of farmland in Oriental Mindoro, central Philippines, to grow corn. South Korean multinationals and provincial governments have also purchased land in Indonesia, Cambodia and Mongolia.

AFRICAN MINERAL RESOURCES AND ARABLE LAND

Africa is generously endowed with huge mineral resources and it is largely believed that three-quarters of its reserves are yet to be discovered. The continent commands nearly 40 per cent of the world's gold and over 85 per cent of platinum and chromium. It is also the top producer of vanadium, cobalt, diamonds, and chrome, and its iron ore reserves are astoundingly high. African mineral reserves rank first or second for bauxite, cobalt, diamonds, phosphate rocks, PGM, vermiculite, and zirconium.

Many other minerals are also present in sizeable quantity. The 2005 share of world production from African soil was as follows: bauxite 9 per cent, aluminium 5 per cent, chromite 44 per cent; cobalt 57 per cent, copper 5 per cent, iron ore 4 per cent; steel 2 per cent; lead (Pb) 3 per cent, manganese 39 per cent, zinc 2 per cent; cement 4 per cent, natural diamond 46 per cent, graphite 2 per cent, phosphate rock 31 per cent, coal 5 per cent, mineral fuels (including coal and petroleum) 13 per cent, uranium 16 per cent. Moreover, it has 60 per cent of the world's uncultivated arable land inviting attention of countries like China and India to meet their burgeoning food and other crop demands.

The era of imperialism came to an end with World War II but the practice of neo-colonialism remains widespread. Economic neo-colonialism can be viewed as the global economic system comprising wealthy countries at the centre and poor countries at the periphery. Economic neo-colonialism extracts the human and natural resources of peripheral or poor countries to flow to the economies of the wealthy countries at the centre of the global economic system.

Decolonised countries having democratic regimes are still dependent on European nations one way or the other. French Areva extracts 40 per cent of its uranium from its former colony Niger. France heavily relies on nuclear energy for 75 per cent of its electricity generation. Niger is expected to become the world's second largest uranium producer within a few years. France, in fact, wants to gain a strong foothold in West Africa in order to secure consistent supply of natural resources, which was seen when it sent its troops to Mali in early 2013, though the pretext was fighting radical Islamist groups. Moreover, military presence evokes fear among other competitors here, chiefly China.

In terms of the ranking of Africa's major trading partners as of 2011, China leads the pack with trade worth $107 billion followed by India with $32 billion. Brazil with trade of $20 billion occupies the third slot and at the fourth place is Russia at $3.5 billion.

The fast and furious growth of India has resulted in rapid depletion of its resources, and to grow further, more and more resources are required. Hence, many Indian companies are seeking fresh opportunities and new avenues in the African continent. In the past few years, many industrial giants have made significant investments in Africa, prominent among them being Minerals and Metals Trading Corporation (MMTC), Vedanta, Insa Coal Holding (Action Group), National Small Industries Corporation (NSIC), Sahara Holdings, Sterlite Technologies, Dabur, Emami, Godrej, Osho South Africa Coal Mining, L&T Africa Initiative, Jindal Africa Investments and Apollo Tyres. Indian rose company Karuturi Global acquired massive land block of around 300,000 hectare (ha) in Ethiopia as well as Kenya. India has announced ambitious plans to invest $70 billion in Africa.

In addition to countries like China, India, Brazil and Russia; Turkey and Iran too wish to augment their business activities in Africa. China clearly wants a lion's share here also. It is wooing Africa through soft power by opening medical care centres, educational institutes, according scholarships and providing long-term interest-free loans.

The year 2014 saw China's fresh return to African mining after a hiatus of nearly two years seeking out copper, iron ore and uranium deposits. Since the beginning of the year, deals started to trickle in: China National Nuclear Corporation took a large stake in one of Africa's largest uranium mines in Namibia for nearly $200 m and China National Gold was in talks to buy a copper mine in Congo. China surpassed the US as Africa's main trading partner in 2009. China is fast emerging as a key player here and fiercely challenging established western dominance. It is actively participating in different business activities. In Mozambique it is extending partnership in agricultural development whereas in Angola, Ghana and Congo it has extended loans for large infrastructure development. Further activities extend to building alumina refinery in Ghana as well as oil refineries in Chad and Niger.

China's African policy is largely governed by a seven-page policy paper issued in January 2006 by the Ministry of Foreign Affairs which proclaims that China does not seek to intervene in the internal affairs of independent states, and is only interested in business. The document pledges to facilitate access of African commodities to Chinese markets, while helping Chinese companies roughly numbered at over one million to operate there, through the provision of preferential loans and buyer credits. Infrastructure and resources are specifically mentioned as key sectors for cooperation.

However, it appears that the policy of non-interference underwent a change in 2013, the very first year of Chinese President Xi Jinping's reign and has shown several new trends that illustrate Beijing's evolving priorities and strategies in the African continent. Most strikingly, China has assertively enhanced its direct involvement in Africa's security affairs. In an unprecedented move, China dispatched around 200 combat troops from the People's Liberation Army (PLA) Special Force to the United Nations peacekeeping mission in Mali in early 2014. In the past, China sent only non-combat staff. This new development suggests Beijing's rising interests, enhanced commitment and direct role in maintaining peace and security in Africa.

In another unprecedented yet more surprising move, China got involved in an open intervention in the South Sudan conflict through direct mediation. In 2013, China's special envoy for African affairs Zhong Jianhua paid several visits to Africa to coordinate positions and mediate the South Sudan issue. Then, in January 2014, in a rare overt political intervention, Chinese Foreign Minister Wang Yi publicly called for an immediate end of hostilities in South Sudan. At Ethiopia's invitation, Wang Yi travelled to Ethiopian capital Addis Ababa to meet with rebel and government delegations. He openly urged 'immediate cessation of hostilities and violence' and publicly called for the international powers to back the Ethiopian-led mediation efforts. Given China's considerable oil stake in the unstable South Sudan (China imported nearly 14 million barrels of oil from South Sudan in the first 10 months of 2013, twice that from Nigeria), a strong impression is gaining ground that China is gradually abandoning its long-term 'non-interference' policy in Africa purely to protect its overseas economic interests.

China produces over 15 per cent of the world's GDP, is the world's prime exporter and has an insatiable appetite for food, territory, minerals and energy to fuel its industrial machine. This is why it has developed a large footprint on the African continent.

In the past couple of years, Brazil too has begun to invest aggressively in Africa. Former president Luíz Inacio Lula da Silva shifted his country's focus to economics, trade and investments, as well as agriculture, oil, mining, infrastructure, health, and science and technology. President Dilma Rousseff has been pursuing the same more rigorously, announcing a new African initiative.

Brazil's major trading partners in Africa are Nigeria (32 per cent), Angola (16 per cent), Algeria (12 per cent), South Africa (10 per cent), and Libya (7 per cent). Together, these countries make up 77 per cent of its total trade with the continent. The bulk of Brazil's imports from Nigeria, Angola and Algeria are of oil. Vale of Brazil, the world's second largest iron producer, boasts $7.7 billion worth of investments in nine African countries. Vale is actively involved in mining exploration—mostly iron ore and copper—in Angola, DRC, Gabon, Ghana, Guinea, Liberia, Mozambique, South Africa, and Zambia.

Brazil's exports to Africa jumped from $2.4 billion in 2002 to $12.2 billion in 2011. Overall, trade with Africa increased from $4.3 billion in 2000 to $27.6 billion in 2011. Brazilian companies with interests in Africa, such as oil giant Petrobras, mining conglomerate Vale and construction firm Odebrecht have pledged to increase investment in Africa.

Petrobras has invested in several offshore oil exploration

projects in Nigeria, Angola, Libya, Namibia, and Tanzania. The company's production in Africa (57,444 boe per day) corresponds to 24 per cent of its international production. Angola is the recipient of its fourth biggest international investment, worth $900 million for the period 2009-2013. Investments in Nigeria are estimated at $2 billion over the same period.

Brazil's leading investment bank Banco BTG Pactual also announced a $1 billion Africa investment fund, the biggest in the world, with capital collected in the country. Brazilian Vale's coal mine in Mozambique is its biggest operation outside Brazil. Odebrecht has been building things in Africa since the 1980s. Earlier, it was involved in construction of the vast Capanda dam in Angola. It erected the country's first shopping mall in the capital, Luanda. In Ghana, where demand for homes is so fierce that tenants have to pay up to two years rent in advance, OAS, a contractor of Camargo Corrêa, a big conglomerate, is putting up social housing. Andrade Gutierrez, another construction firm, works on everything from ports to housing and sanitation projects in Angola, Algeria, Congo and Guinea. Petrobras, Brazil's state-owned oil behemoth is already pumping oil in Angola and Nigeria and is on the hunt for more in Benin, Gabon, Libya, Nigeria and Tanzania.

From the above narration, it is quite evident that a new stage has been set in human history for resource exploitation. Most of the planet's easily-accessed deposits have now either been depleted or are already in the process of exploitation/ over-exploitation. Newer territories are being acquired and the quest has gone to hellish geography of the Arctic, virgin Senkaku and under-exploited Africa and central Asia that are now being invaded with enormous vigour. Opting to operate in the Arctic would result in multi-dimensional ramifications. The companies would face much greater risks than they do elsewhere. Any chance oil spill, however provoked, would be truly catastrophic, as the region's flora and fauna are already highly stressed and therefore more vulnerable to toxic chemical effects than their brethren further south, and because it would be much harder to mount a rescue operation here than in places like the Gulf of Mexico, where numerous boats and crews are available to help contain the disaster.

The desperate pursuit of the world's resources is also behind the recent flare-up of tensions in the East and South China Sea, involving China, Japan, Taiwan, Vietnam, and the Philippines, between China and the US, between Russia and EU, and so on and so forth. Fresh disputes will soon follow suit!

Growing competition among rich nations to exploit energy and mineral resources will wreak havoc upon highly fragile ecosystem of virgin islands, Arctic ice sheet and forest resources globally. This will engender severe environmental hazards along with threats to the traditional lifestyles of indigenous people and ultimately produce increased tension and hostility among the major consuming countries, leading to a new Cold War like atmosphere.

Quest for Renewable Resources

IN THE annals of evolution of alternative energy, 1861 was a truly momentous year. That was when Augustin Mouchot invented the active steam engine powered by the sun, paving the way for subsequent inventions in the alternative energy domain.

What is alternative energy? In layman's term, fossil fuel is the remains of animals and plants that lived up to millions of years ago. This natural resource is non-renewable. In earlier times, mankind heavily relied on wood for all its energy needs. In the Middle Ages, coal became an alternative to wood, and was used to heat houses and cook food. In the early nineteenth century, petroleum became an alternative to whale oil, which was widely used for lighting lamps and resulted in serious destruction of whales in the oceans. The dawn of the twentieth century was marked with the appearance of ethanol as the alternative to petrol. In fact, any new fuel discovery by humans becomes an alternative to the old ones at some point of time.

Nowadays, the most plausible definition of alternative energy could be rendered as that derived from sources which do not deplete or exhaust natural resources and do not harm the environment. Thus, alternative energy in this era constitutes solar, wind, geothermal or other renewable energy that is capable of replacing or supplementing traditional fossil-fuel sources such as coal, oil and natural gas.

Nuclear energy is neither renewable nor alternative energy as the main fuel source. Uranium has limited availability of 100 years or so and also inflicts severe environmental damage in the form of harmful radiation.

Today, the quest for alternative energy is more relevant, logical and inevitable amidst soaring crude oil prices hovering at $100-150 a barrel, rising Global Average Temperature (GAT), ocean acidification, land degradation by over-extraction, limited availability of coal, gas and gasoline and above all atmospheric CO_2 concentration not plunging below 400 parts per million (ppm).

GEOTHERMAL ENERGY: A BOON FOR ACTIVE VOLCANO REGIONS

Geothermal energy is a resource that has been used by mankind for washing and healing through the ages. In the preceding century, it became a source for electricity production and was used directly, apart from bathing or washing, for heating of houses, greenhouse heating and aquaculture. The quest for this alternative source of energy has now intensified and many countries have joined the race as access to clean energy at affordable prices has become a key issue to improve the standard of living. Moreover, at present, around two billion people, a third of the world's population, have no access to modern energy sources. Furthermore, the world population is expected to double by 2100, which will double the world's energy consumption.

Traditionally, geothermal energy use has been very site specific, normally concentrated in volcanically active regions of the world. However, recent technological advancements like enhanced geothermal systems (EGS) could significantly widen the scope of the sector and make geothermal energy one of the major sources of energy in the twenty-first century. It is possible to drill down as deep as four kilometres to retrieve hot fluids and steam for power production. This is a highly complex task.

There are currently two dozen countries around the world that generate electrical power from geothermal power plants. The international geothermal power market is booming, growing at a sustained rate of 4-5 per cent. Almost 700 geothermal projects are now under development in some 76 countries. Many countries, anticipating the threats caused by Climate Change, realise the value of geothermal power as a base-load and sometimes flexible source of renewable energy. These counties are on every continent and range from small island nations to large developed economies.

Current International Scenario

As of 2019, geothermal electricity is being generated in 24 countries across the globe. Production has significantly grown from 9,732 megawatt (mw) in 2007 to 14,500 mw in 2018.

Below is the list of Top 10 geothermal countries at the year-end in 2018.

United States:	3,639 mw – with an additional 48 mw just added before end-2018
Indonesia:	1,948 mw – short of 95 mw planned to be added before year-end
Philippines:	1,868 mw – projects were to come online in 2020
Turkey:	1,347 mw – a lot of capacity came online just before the year-end to take advantage of favourable feed-in-tariffs
New Zealand:	1,005 mw – with 25 MW in commission just before the year-end, the country just managed to get back into the 1 GW Geothermal Country Club
Mexico:	951 mw – with an exciting new incentive and support scheme, things might be moving further in the coming years
Italy:	944 mw – with the current political climate, this number might not change much soon
Iceland:	755 mw – adding another 45 mw in early 2019
Kenya:	676 mw – with the pipeline of projects, the country will jump up in the ranking soon
Japan:	542 mw – small-scale developments but overall not much progress.

Other countries represent an installed power generation capacity of 925 mw, which brought the total installed geothermal power generation capacity at the end of the year 2018 to 14,600 mw.

Countries in Race

Significant growth is expected in the global geothermal power space over the next few years as countries in eastern Africa push projects toward completion. Kenya and Ethiopia are making rapid strides by building power plants greater than 100 mw whereas the average size of a geothermal power plant in the United States is about 25 mw. Steady growth continues in South Pacific despite setbacks from typhoon Haiyan in 2013. Plants that were damaged were quickly repaired and most normal geothermal activities resumed in the Philippines by February 2014.

Some countries are growing so quickly there could be a time in the near future when the US may no longer remain the world leader in geothermal power, despite its vast geothermal resources.

Meanwhile, Indonesia has 4,400 mw of capacity addition in the pipeline. It is unlikely all of these projects will come to fruition quickly since Indonesia is struggling with a number of bureaucratic barriers hindering development, such as land acquisition problems. There could be a time in the near future where countries passionate about geothermal power and comparable geothermal reservoirs are close competitors for the title of top producing nation.

A few countries in Central and South America such as Argentina, Chile, Columbia and Honduras have significant amounts of geothermal potential; however, these countries are still in the early stages of exploring and identifying their resources. In the coming years, it is expected that announcements for planned capacity addition in these countries will likely become more frequent as geothermal resources can be identified for electricity generation. Chile is actively developing 50 early-stage projects and prospects.

Small island nations such as Nevis and St Lucia are dedicated to powering a substantial portion of their economies from geothermal power. However, these small island nations often go unnoticed because their economies and therefore their projects barely register on the global map. That isn't a reason to doubt the progress they are making. Investment is quickly pouring into island nation projects and as per the annual US and Global Geothermal Power Production Report (April 2014). Meanwhile, more experienced island nation geothermal markets like Costa Rica have found funding and are building new geothermal plants after several years of stagnation.

Some established geothermal markets that have not seen new growth in many years are revisiting geothermal development. For example, Mexico passed legislation at the start of 2014, opening the electricity market to private investors. In Japan, feed-in-tariffs are revitalising the geothermal power market, particularly for small-sized projects.

Looking closely at projects in the pipeline, global capacity could be at 14,000 mw in 2020, amply attested by several hundred megawatts of new geothermal power becoming operational per year. The leader in terms of megawatts under construction is Indonesia, with roughly 425 mw, followed by Kenya with over 296 mw of geothermal power currently under construction. Some other countries of note in this context are the Philippines with 110 mw, Iceland with 260 mw, New Zealand with 166 mw, and the US with 178 mw of geothermal power currently under construction. Germany is closely following the US with the most new power plants under construction with eight new power plants and 47 mw. However, they are smaller projects ranging from 1-6 mw each. Many new power plants came online in the US, the Philippines, Mexico, New Zealand, Germany, Kenya, Australia, and Turkey in 2013. Kenya, Turkey, Ethiopia, Costa Rica, and Germany are quickly developing geothermal power infrastructure. In addition, it is likely within the next decade or so the Philippines or Indonesia could roughly equal the US in capacity. By looking at projects in the pipeline, other smaller countries are likely to become more established geothermal power markets.

The United States, Indonesia, Philippines, Turkey and New Zealand constitute the Top 5 club with geothermal power projects or prospects under development. These countries have significantly added geothermal in their energy basket. During 2008-18, Turkey made the highest progress by adding 1,312 mw of geothermal power followed by Indonesia (956 mw), US (599 mw), Kenya (509 mw) and New Zealand (421 mw).

Infographic 3.1: Top 5 Established Leaders in terms of Installed Geothermal Capacity (>950 mw) by 2016

*Capacity in Megawatt (mw)

** Source: GEA (Geothermal Energy Association) Report 2016

Infographic 3.2: Countries having Installed Capacity between 200 mw and 950 mw by 2016

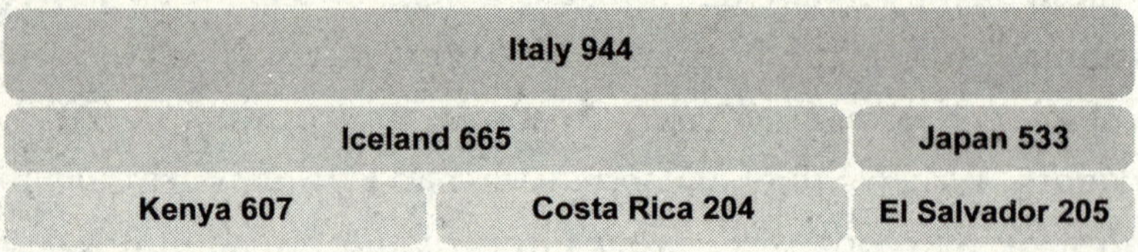

*Capacity in megawatt (mw)

** Source: GEA (Geothermal Energy Association) Report 2016

Emergence of Philippines and Indonesia

Both the Philippines and Indonesia have made rapid strides in the geothermal energy space and their rise has been spectacular. Though the Philippines and Indonesia now represent the second and third largest installed bases of geothermal power in the world, yet the region has barely scratched the surface of potential geothermal resources. Indonesia has ambitiously targeted around 4 gw of new geothermal capacity by the end of 2014. Meanwhile, the Philippines targeted over 1.4 gw of increased capacity as part of its long-term renewable energy plan looking to 2030, with 1,020 mw of new installations expected to come online between 2015 and 2020. The incredible growth of the Indonesian energy market has been fuelled by a growing energy-hungry population. Since 1985, the Indonesian energy market has seen per annum growth of 12 per cent even during the regional recession in 1998-2000. Estimates vary on how much geothermal energy will be developed in the next 5-10 years, but conservative estimates see approximately 2-4 gw with an average cost of $4 billion per gw. Apart from huge domestic demand and ability to create additional job opportunities for its people, Indonesia can tap the electricity export markets of energy-thirsty neighbours such as Malaysia and Singapore.

The Indonesian government is concerned about its huge diesel fuel subsidy bill, which currently stands at $9 bn per annum and is keen to redirect some of this into seeding clean energy projects. Although 9.7 c/kwhr is the much quoted ceiling for Power Purchase Agreements (PPAs), it can be increased depending on the location of the project by 2-3 c/kwhr outside of the main islands of Java and Sumatra. In fact, the excellent resources available make projects feasible at even 8.8 c/kwhr and lower.

Entry of Big Global Players

Origin Energy of Australia, Tata Power of India and British Petroleum are commited to develop the Indonesian geothermal industry. These companies together with Total, Chevron and EDC Philippines all have the resources and balance sheets to finance their own projects.

With numerous projects in the development phase and new tenders yet to be won, there are significant opportunities for geothermal stakeholders to capitalise on these markets. GeoPower Indonesia and Philippines has been accelerating the pace of progress in the realm of geothermal energy in these regions and bringing together policy makers with local and international geothermal companies

Icelandic Ventures in Geothermal Energy: An Inspiring Story

During the course of the twentieth century, Iceland went from being one of Europe's poorest countries, dependent upon peat and imported coal for its energy, to a country with a high standard of living where practically all stationary energy is derived from renewable resources. In 2011, roughly 84 per cent of primary energy use in Iceland came from indigenous renewable resources, out of which geothermal share was 66 per cent. Iceland, no doubt, has been a pioneer in the use of geothermal energy for space heating. Generating electricity with geothermal energy has increased significantly in recent years. Geothermal power facilities currently generate 25 per cent of the country's total electricity production.

Direct Use of Geothermal Resources in Iceland

Iceland is well known to be a world leader in the use of geothermal heating. After World War II, Orkustofnun (National Energy Authority of Iceland) carried out intensive research and development, which led to the use of geothermal resources for heating of households, Today, about nine out of ten households are heated with geothermal energy. Space heating is the largest component in the direct use of geothermal energy in Iceland. In the year 2011, the total use of geothermal energy was 422 petajoule (pj), with space heating accounting for 45 per cent.

The Iceland Deep Drilling Project (IDDP) is a long-term study of high-temperature hydrothermal systems in Iceland. It is a collaborative effort by a consortium of Icelandic power companies and the government, formed to determine if utilising super critical geothermal fluids would improve the economics of power production from geothermal fields.

Over the next several years, the IDDP expects to drill and test a series of boreholes that will penetrate super critical zones believed to be present beneath three currently exploited geothermal fields in Iceland. This will require drilling to a depth of about 5 km in order to reach hydrothermal fluids at 450-600°C temperatures.

A feasibility study completed in 2003 indicates that relative to the output from conventional geothermal wells, which are 2.5 km deep, a ten-fold increase in power output per well could result if fluid is produced from reservoirs hotter than 450°C. A typical 2.5 km-deep geothermal well in Iceland yields power equivalent to approximately 5 mwe. Assuming a similar volumetric inflow rate of steam, an IDDP well tapping a super critical reservoir at temperatures above 450°C and at a pressure of 23-26 MPa may be expected to yield around 50 mwe.

ENERGY BLOWING IN THE WIND

Mankind has been using wind energy since antiquity in the form of windmills and watermills that were the source of motive power for a number of mechanical applications, some of which are used even today. Humans have been using wind energy in their daily work for over four millennia. Sails revolutionised seafaring, which no longer had to make do with muscle power. In 1700 BC, King Hammurabi of Babylon used wind-powered scoops to irrigate Mesopotamia. Apart from pumps for irrigation or drainage, windmills were mostly used to grind grain. Thus, we still speak of 'windmills' today, even when we are talking about machines that do not actually grind, such as sawmills and hammer mills.

A sudden spurt is being witnessed in harnessing wind power with over a hundred countries now vying for this non-conventional energy derived from munificent Nature. Let's cast a glance at statistics to understand the phenomenon.

By the end of 2013, World Wind Energy capacity reached 318,529 mw (with addition of 35,550 mw new capacity) surpassing previous year's capacity of 282,275 mw. Wind power now contributing and meeting close to 4 per cent of global electricity demand. No less than 103 countries are today commercially exploiting wind power. Here also, China occupies the forefront with new wind power capacity of 16,000 mw and a total capacity of 91,324 mw. However, the US market saw a dramatic slump and installed only 1 gw in 2013 after installing 13 gw in 2012.

Total installed onshore wind power capacity in Asia-Pacific in 2018 was 256 gw. Leading countries in the region for onshore wind turbines installations were China (21.2 gw) and India (2.2 gw). The world capacity is 650.8 gw, 59.7 gw added in 2019, according to the statistics portal https://wwindea.org/blog/category/statistics/. The most dynamic markets with highest growth rates can still be found in Latin America and Eastern Europe as well as for the first time in Africa, where Morocco showed a growth of 70 per cent, the second highest growth rate of all countries, just behind Chile (76 per cent).

In some European countries, wind power has very high shares in electricity supply. Wind energy has actually become the largest source of electricity in Denmark (34 per cent share) and Spain (21 per cent). Countries like Portugal (over 20 per cent), Ireland (over 16 per cent) and Germany (9 per cent) have reached high proportions. By end-2013, 74 gw of offshore wind energy was installed, and close to one million small wind turbines. In 2013 the total global wind power generation was 318,914 mw which rose to 596,556 mw by the end of 2018. By 2020, global wind capacity was expected to cross 700,000 mw.

Rising Market of Small Wind Turbines

The world market for small wind has recorded sustained growth over the period. China, the US and UK now account for 80 per cent of the small wind world market. As of the end of 2012, a cumulative total of at least 806,000 small wind turbines were installed all over the world. This is an increase of 10 per cent compared with the previous year, when 730,000 units were registered.

In this segment, most of the growth and rivalry happens among three countries: China, US and UK. This situation is a clear indication that the world market for small wind turbines is still in its infancy. In most countries, one can find only a handful of small wind turbines and no mass production. Better policies are imperative for making small wind a success all over the world.

The recorded small wind capacity installed worldwide reached more than 678 mw by the end of 2012. This was a growth of 18 per cent compared with 2011, when 576 mw were registered. In 2011, the growth rate was still at 21 per cent. China accounts for 39 per cent of global capacity, the US for 31 per cent and UK for 9.4 per cent. These three leading markets—China, US and UK—together installed around 89 mw of new capacity in the year 2012 (80 per cent of the world market), a capacity increase of 16 per cent, slightly below the global growth rate.

Global Players Harnessing Wind Power

Around five dozen prominent companies have been active in over 100 countries in the global race for exploiting more and more wind energy. Some mergers and acquisitions have been witnessed in the preceding years. The sector is likely to witness hectic activity in times to come. The sector is set to consolidate further.

The Top 10 wind turbine manufacturers by annual market share in 2018 are:

I. Vestas (Denmark): 10.09 gw
II. Goldwind (China): 6.66 gw
III. GE Wind Energy (US): 4.96 gw
IV. SGRE (Siemens Gamesa Renewable Energy): 4.08 gw
V. Envision (China): 3.28 gw
VI. Enercon (Germany): 2.53 gw
VII. Ming Yang (China): 2.44 gw
VIII. Nordex (Germany): 2.43 gw
IX. Guodian United Power (China): 1.29 gw
X. Windey (China): 0.94 gw

Denmark's Vestas is no doubt the undisputed king and market leader with nearly 22 per cent share in its slice of the pie up from 16 per cent in 2017. China's Goldwind rose from third to second slot mainly lifted by its performance in China where it captured a third of the market. Five out of Top 10 best wind power producer companies are now from China. Other prominent players either come from the North America or Europe. Here is the alphabetical list of renowned players in the realm of wind power technology:

Acciona Energy (Spain)
Aeolus Windpower (China)
Alstom Ecotècnia (Spain)
A Power Energy Systems (China)
Aeronautica Windpower (US)
AREVA (France, purchased Multibrid of Germany in June 2010)
Argosy Wind Power (USA)
Bornay (Spain)

BraunWindturbinen (Germany)
Clipper Windpower (USA)
DyoCore SolAir (USA)
Daewoo Shipbuilding and Marine Engineering (Korea)
DeWind (Germany/USA, bought by Daewoo Shipbuilding and Marine Engineering in 2009)
Doosan (Korea)
Ecotechnia (Spain)
Enercon (Germany)
Gamesa Eolica (Spain)
General Electric (USA)
Goldwind (China)
Hanjin (Korea)
Hyosung (Korea)
Hyundai Heavy Industries (Korea)
INVAP (Argentina)
Koncar (Croatia)
Lagerwey Wind (The Netherlands)
Leitner (Italy)
LM Wind Power (Denmark)
Ming Yang (China)
Mitsubishi Heavy Industries (Japan)
NEG Micon (now part of Vestas)
Nordex (Germany)
Nordic Windpower (USA)
Northern Power Systems (USA)
PacWind (USA)
PitchWind Systems (Sweden)
Quietrevolution (United Kingdom)
Raum Energy Inc. (Canada)

ReDriven Power Inc. (Canada)
Repower (Germany, nearly 90 per cent owned by Suzlon)
Samsung Heavy Industries (Korea)
SANY (China)
Scanwind (Norway, bought by General Electric in 2009)
Schuler (Germany)
Siemens Wind Power (Denmark/Germany)
Sinovel (China)
Sonkyo Energy (Spain)
Southwest Windpower (USA)
Soyut Wind (Turkey)
SRC Green Power (India)
STX Windpower (The Netherlands)
Suzlon (India)
Urban Green Energy (USA)
Vestas (Denmark)
WindAid (Peru)
WinWinD (Finland)
Windflow (New Zealand).

Competing States and Respective Share

Some states are fortunate enough to possess all the requisite parameters such as technology, unobstructed wind flow for the greater part of the year, vast empty space and other favourable topographic features whereas others are not that lucky. This stark contrast has rendered only five nations with the lion's share: China, US, Germany, Spain and India alone today occupy nearly three-fourths (72.23 per cent) of the total world capacity. The remaining 27.77 per cent is shared among 98 competing nations.

Total installed capacity of the world was reckoned as gw 318,137 mw by the end of 2013. The Top 10 countries were: China (91,424 mw), United States (61,091), Germany (34,250), Spain (22.959), India (20,150), United Kingdom (10,531), Italy (8,552), France (8,254), Canada (7,803), Denmark (4,772) and the rest of world (48,351 mw).

Although by end of 2019, worldwide wind energy did reach 650,8 gw capacity, the predicted 700 gw is, however, unlikely to happen due to a production slowdown because of the corona virus pandemic worldwide.

India's Finger in the Pie

With total wind installed capacity of around 35,815 mw as of 30 April 2019, India is the world's fourth largest country in terms of total wind installations after China, the US and Germany. Wind power has become one of the key renewable energy sources for power generation in India, contributing a share of at least 6-7 per cent to the country's electricity generation mix at present. Since India has declared the renewable energy target of 175 gw by 2022, the wind power programme has gained further prominence in this scheme.

It is estimated that with the current level of technology, the 'onshore' potential for utilisation of wind energy for electricity generation is to the tune of 65,000 mw. India is blessed with 7,517 km of coastline, with territorial waters extending up to 12 nautical miles (nm) into the sea. This enables India to exploit both on-land and offshore wind energy potential. The unexploited resource availability has the tremendous potential to sustain the growth of the wind energy sector in India in years to come. Many major world companies are pouring into the fast evolving wind energy market in India: Enercon, Vestas, GE Wind, and Gamesa have already opened up their establishments across various cities in India.

Numerous wind energy companies (domestic as well as overseas) in India are active and making a significant contribution to the green energy sector. These are Suzlon Energy, Regen Powertech, RRB Energy, SRC Green Power, Lietnar Shriram, Gamesa, WinWind, Chiranjjeevi Wind Energy, Kenersys and Pioneer Wincon. At present, Suzlon is leading the pack.

In 2006, when the Ministry of New and Renewable Energy (MNRE) was instituted, India became the first country in the world to create an exclusive ministry for the development of the renewable energy sector. The main aim of the ministry is to develop and deploy new and renewable energy to achieve the national goal of energy security.

Several Indian states like Karnataka, Andhra Pradesh and Tamil Nadu have also come up with renewable energy policies and the Indian government has been actively pursuing and supporting wind energy programmes. A new estimation of wind power potential at 100 m hub-height of prospective Indian States/UTs is estimated at 302 gw.

State	Rank I*	Rank II*	Rank III*	Total
Andaman & Nicobar	4.12	3.43	0.88	8.43
Andhra Pradesh	22525.5	20538.1	1165	44228.6
Chhattisgarh	3.24	57.03	16.31	76.59
Goa	0	0.08	0.76	0.84

Gujarat	52287.59	32037.83	105.09	84431.33
Karnataka	15202.36	39802.59	852.4	55857.36
Kerala	332.63	1102.56	264.38	1699.56
Lakshadweep	3.5	3.4	0.77	7.67
Madhya Pradesh	2216.39	8258.55	8.93	10483.88
Maharashtra	31154.76	13747.43	492.15	45394.34
Odisha	1666.2	1267.06	160.22	3093.47
Puducherry	69.43	79	4.4	152.83
Rajasthan	15414.91	3342.62	12.96	18770.49
Tamil Nadu	11251.48	22153.34	394.82	33799.65
Telangana	887.43	3347.52	9.34	4244.29
West Bengal	0.03	2.04	0.01	2.08
Total in MW	153019.59	145742.59	3489.31	302251.49
Total in GW	**153**	**146**	**3**	**302**

*Source- NIWE India

* Rank I – Wasteland, * Rank II – Cultivable Land, * Rank III – Forest Land

Reaching Deep Seas: Deep Offshore Wind Farms

Deep offshore wind turbines are in a nascent stage of development, with Europe taking the lead. By the end of 2012, two full-scale grid-connected offshore wind turbines on floating substructures were installed: Hywind (Norway) in 2009 and Winfloat (Portugal) in 2012, located in the North Sea and Atlantic. Hywind is developed by Statoil, with a 2.3 mw Siemens turbine. Installed in Norway in 2009, it is the first large scale floating wind structure of the world.

Windfloat, the second large-scale floating system, was installed off the Portuguese coast in 2011 and commenced commercial production in 2012. It was developed by Principle Power and EDP with a 2 mw Vestas wind turbine.

Seven experimental floating sub-structures (four in Europe, two in Japan and one in the US) are in a testing phase: Sea Twirl, SWAY, Blue H and Poseidon in Europe, Kabashima Island concept and WindLens in Japan and DeepC wind floating turbine in the US.

The offshore wind industry, though in its infancy, has huge potential for cost reduction and technological innovation. Europe has been increasingly developing deep offshore designs for water depths of over 50 m to unlock market potential. In addition, two full-scale deep offshore turbines at the end of 2012, there are three grid-connected experimental floating substructures and thirty-five deep water designs under development worldwide.

Of all forty projects identified, either grid connected systems or under development, twenty-seven are located in nine European countries: Denmark, France, Germany, the Netherlands, Norway, Portugal, Spain, Sweden and the UK. Four are in the US and nine in Japan.

Environmental Implications of Offshore Wind Energy

Wind energy is universally acknowledged as the most efficient, clean and green energy but it is also fraught with some implications which have recently been recognised. Some concerns about the environmental impacts of offshore wind turbines are enumerated below:

The foremost concern is the impact of the increased noise

on marine life. Noise is produced during the construction and installation of offshore wind farms, from increased boat activity in the area and procedures such as pile-driving. The sound levels from pile-driving, when the turbine is hammered to the seabed, are particularly high. This is potentially harmful to marine species and construed as the greatest concern to aquatic mammal species such as endangered whales.

The noise and vibration of construction and operation of the wind turbines can be damaging to fish and other marine species. The effects of noise may be immediately fatal, cause injuries, or result in short or longer term avoidance of the area depending on the frequency and decibel level of the sounds.

There is potential risk of death to birds and bats from direct collisions with the rotors and the pressure effects of wind vortices. There is also a risk of displacement from the area causing changes in migration routes and loss of quality habitat.

Construction activities at the wind power site and the installation of undersea cables to transmit the energy to shore can have direct effects on the seabed and sediments, which can affect the abundance and diversity of benthic organisms (those that live at greater depths).

Disturbance of the seafloor may also increase turbidity, which could affect plankton in the water column.

Considerate site selection of wind turbines, mitigation strategies such as the presence of marine mammal observers and a 'soft-start' to pile-driving, where hammering starts softly so the sound slowly increases, and alternative types of foundation that do not require pile-driving can help reduce these impacts to some extent.

There are also potentially some environmental benefits of

offshore wind farms. The turbines may act as 'artificial reefs' and increase biological productivity in the vicinity. The presence of hard structures can provide habitat for barnacles, sponges, and other invertebrates. Fishing will be prohibited around the turbines for safety, which may increase fish abundance locally. These processes can consequently result in attracting predators higher up the food chain.

Predators like Petrel Southern Giant, Frigate bird, Black-backed Gull, Polar Skua, long-tailed Jaegar, red-breasted Merganser, Pelican, blue heron, Heermann's Gull and Caspian Tern would throng in large numbers and receive plentiful food in the form of burgeoning fish, thus enriching and enhancing the biodiversity of the whole area which is dwindling globally at an alarming rate and render the food chain more robust and sustainable!

TAPPING OCEAN WAVE ENERGY

Human endeavour is now gradually shifting to extract oceanic wave energy and convert it into electricity. According to the World Energy Council (WEC), two terawatts (tw) of energy, or the equivalent of twice the world's electricity production, could be harvested from the world's oceans.

Then there is wave energy! It's quite logical to seek this new vista as wave power is the most predictable and dependable form of renewable energy. Moreover, viable and feasible wave power sites are available around the world, especially close to population centres. There are many high wave power sites close to densely populated areas around the continents. It is estimated that 1 million gw hours of wave energy hits Australian shores annually and that 25 per cent of the UK's current power usage could be supplied by harvesting its wave resource.

In 2011, consistent with other industry forecasts, the UK Carbon Trust estimated that the Global Marine Energy sector could be worth $760 billion by 2050 and could support 68,000 jobs in the British marine energy sector alone. Industry estimates put annual marine energy revenues at nearly $100 billion by 2025. Wave energy has significant global potential with the US, North and South America, West Europe, Japan, South Africa, Australia and New Zealand among some of the best wave energy sites around the world.

Wave energy flows in the direction of wave propagation and is measured as the amount of power (in kw) contained in each linear metre of wave front. Typically, energy above 20 kw/m may be suitable for most wave farm projects and many devices work well above this level.

Evolution of the Concept

The idea of producing electricity from wave energy is not novel. The first ever known patent to use energy from ocean waves dates back to 1799 and was filed by Girard and his son in Paris. A device to tap wave power was developed around 1910 by Bochaux-Praceique to illuminate his home in France. This was perhaps the first oscillating water-column type of wave-energy device. Over 300 patents were filed in the UK alone during the period 1855–1973.

Modern scientific pursuit of wave energy was pioneered by Yoshio Masuda's experiments in the 1940s. He tested several concepts of wave energy devices at sea. The concept of extracting power from the angular motion at the joints of an articulated raft was also proposed by Masuda during the 1950s.

Renewed Global Interest in Wave Power

Hope, disappointment and renewed hope—this has been the story so far for wave energy. Hope was rekindled owing to a number of factors which are equally responsible for the quest of other alternative energy sources. The first surge of interest in wave energy was largely dictated by the oil crisis of 1973. A number of university researchers revisited the potential to produce energy from ocean waves, Stephen Salter's 1974 invention became known as Salter's duck or nodding duck, although it was officially referred to as the Edinburgh duck. In small-scale controlled tests, the duck's curved body could stop 90 per cent of wave motion and convert 90 per cent of that to electricity rendering 81 per cent efficiency.

In the 1980s, as oil prices plunged, wave-energy funding was drastically reduced. Nevertheless, a few first-generation prototypes were evolved and tested on site. In this century, following awareness about Climate Change, atmospheric CO_2 level breaching 400 ppm and crude hovering in the $100 per barrel territory, there is a renewed interest worldwide in wave energy vis-a-vis other alternative energy sources.

Important Players/Developers

Most important companies in the field come from Europe, Australia and the US. Most of them started their operations either in late 1990s or early 2000s.

Companies from the UK include Pelamis Wave Power, AWS Ocean Energy, WaveGen, Offshore Wave Energy, ORECon, Ocean Wave Master, C-Wave, Trident Energy, Ocean Navitas and Aquamarine Power. Wave Star Energy and Wave Dragon are prominent players from Denmark.

Sea Power Pacific and Oceanlinx are important companies from Australia. Other companies of note are Syncwave Energy (Canada), WAVEenergy (Norway), Ocean Power (US), Wavebob (Ireland), Seabased (Sweden) and SDE Energy (Israel).

Technologies for Wave Energy

There are three main types of wave energy technologies:

- Floats and Buoys: This technology deploys floats, buoys or pitching devices to generate electricity using the rise and fall of ocean swells to drive hydraulic pumps.
- Oscillating Water Column (OWC): OWC devices generate electricity at the shore, using the rise and fall of water within a cylindrical shaft. The rising water drives air out of the top of the shaft, powering an air-driven turbine.
- Overtopper: In this technology a tapered channel or overtopping device can be located either on or off shore. They concentrate waves and drive them into an elevated reservoir, where power is then generated using hydropower turbines as the water is released.

The vast majority of recently proposed wave energy projects normally utilise offshore floats, buoys or pitching devices.

Notable Wave Energy Farms

Aguçadoura Wave Farm was the world's first, located 5 km offshore near Povoa de Varzim, north of Porto, Portugal. The farm was designed to use three Pelamis wave energy converters to convert the motion of the ocean surface waves into electricity with 2.25 mw installed capacity.

The farm first generated electricity in July 2008 and was

officially opened on 23 September 2008. The wave farm was shut down two months later owing to the financial collapse of Babcock & Brown. The machines were offsite at this time due to technical problems, and though resolved have not returned to site and were subsequently scrapped in 2011 as the technology had moved on to the P2 variant as supplied to Eon and Scottish Power Renewables. A second phase envisages increasing the installed capacity to 21 mw using 25 Pelamis machines.

Reedsport, Oregon, a commercial wave park on the US west coast is located about four km offshore near Reedsport, Oregon. The first phase of this project is for ten PB150 Power Buoys, or 1.5 mw. The Reedsport wave farm was initially scheduled for installation in spring 2013 but the project came to a halt because of legal and technical issues in August 2013.

United Kingdom

Funding for a 3 mw wave farm in Scotland was announced on 20 February 2007 by the Scottish Executive at a cost of over £4 million, as part of a £13 million funding package for marine power in Scotland. The first of 66 machines was launched in May 2010.

A facility known as Wave Hub has been constructed off the north coast of Cornwall, England, to facilitate wave energy development. The Hub will act as giant extension cable, allowing arrays of devices generating wave energy to be connected to the electricity grid. The Wave Hub will initially allow 20 mw of capacity to be connected, with potential expansion to 40 mw. Four device manufacturers have so far expressed interest in connecting to the Wave Hub. Scientists have calculated that wave energy gathered at Wave Hub will be enough to power up to 7,500 households. The site has the potential to save greenhouse gas emissions of about 300,000 tonnes of carbon dioxide in the next 25 years. Operations were halted due to a technical snag in tether when last reports came in. An American company Gwave was to install a new device but that has been postponed.

Australia

A CETO wave farm off the coast of Western Australia has been operating to prove commercial viability and, after preliminary environmental approval, is poised for further development.

Ocean Power Technologies (OPT Australasia) is developing a wave farm connected to the grid near Portland, Victoria through a 19 mw wave power station. The project has received an Au$66.46 million grant from the Federal Government of Australia.

Oceanlinx will deploy a commercial scale demonstrator off the coast of South Australia at Port MacDonnell before the end of 2013. This device, the greenWAVE, has a rated electrical capacity of 1 mw. This project has been supported by ARENA through the Emerging Renewables Programme. The greenWAVE device is a bottom-standing gravity structure that does not require anchoring or seabed preparation and with no moving parts below the surface of the water.

Constraints in developing wave energy

The world has been desperately trying to harness this fascinating energy for the past 150 years but the desired success has been elusive. There are many constraints, limitations and hurdles. Some of them are being listed below:

Design and Site Constraints: As wave power is scattered and the size of individual waves is limited, all designs have necessarily to be modular. Harnessing wave energy probably will not be done with a few, very large generators. Large-scale use of wave energy would entail use of several hundreds of small generators. Wave power is more energy dense than wind power, but it is still diffuse. Research data shows that even in dense high wave energy areas such as the Pacific Northwest, one can expect energy production rates of about 1.5 mw for every 100 feet of shoreline occupied by generators. By comparison, a large fossil fuel plant of 1,000 mw capacity would occupy only about 200 acres. Installing a similar capacity using onshore wave power would occupy over 12.5 miles of shoreline; and that too in the best areas, like the Pacific Northwest.

Exorbitant Costs: Wave power is currently highly capital intensive and too costly to produce. Between the high costs of devices needed to harness the power of the waves to the expensive efforts behind efficient generation of power from waves, it may be some time before the price of wave power can be brought down and made competitive with thermal and nuclear power.

Specific and Limited Locations: While waves cover virtually every mile of water on the planet, economically accessible wave power is found only in coastal areas. Some areas are better than others and the best resources tend to be found only in the specific regions.

Development in Phases: The development process primarily consists of three phases. It begins with small-scale prototype devices that typically have a low capacity. Successful devices lead on to larger capacity prototypes, at this stage outside funding from government or private investors is possible for the most promising devices. The final stage, representing the culmination of development is the production of full-scale grid connected devices that will, in some cases, be deployed in farm style configurations. Only few prototype wave devices are close to entering the final stage and commercial deployment. **Variable Energy Supply:** The energy supply depends on waves and their intensities, which are variable. Even in the most active wave areas, there are many days with little wave activity. On days that have good wave activity, wave levels can vary. Resolving intermittency problems to attain reliable energy output can double and even triple the cost of power. **Effects on Marine Life**: The effects of wave energy systems on aquatic life are not fully understood so far, and this constitutes one among many bottlenecks. **Requirement of High-Strength Device:** One of the most challenging problems is the construction of devices that can withstand wave attacks over and over again. Designing and building a machine that can last for years despite being attacked by furious pounding waves is a daunting task indeed. The future largely depends upon successful design and development of such sturdy machines.

Technology is still in the evolution stage and incessant efforts are needed for attaining perfection. Though a number of wave energy devices are getting closer to full-scale deployment, the fact remains that little real-world operational experience has been gained till date. Large-scale demonstrations are required in order to test survivability and efficiency issues that have not yet been resolved. It is difficult to assess potential of a system until it is tested in its final state.

A small fraction of all wave energy concepts are realisable at a commercial level. Drawing together resources will ensure that the devices that do progress stand the best possible chance of succeeding. The dominance of Small and Medium Enterprises (SMEs) and the stark absence of giant players in the sector is a barrier to development. The have limited resources at their disposal, which constrains progress and development. These small companies are, in most cases, unwilling to collaborate as they naturally wish to protect their own investment.

SOLAR ENERGY

We have discussed geothermal, wind, and wave energy so far; now let us delve into another alternative energy source: solar energy. Solar energy is the energy given out by the sun on a bright, sunny day by means of sunlight which gives off approximately 1,000 watts of energy per square metre of open surface.

Solar is the world's most abundant energy source. Enough sunlight reaches us every 15 minutes to power the whole world for an entire year. There is now technology available to tap into this energy for lighting up our homes and offices. Solar Electricity uses photovoltaic (PV) cells. These PV cells as a group are electrically connected and packaged into a frame (commonly known as a solar panel) which can be grouped into larger solar arrays.

There are pros and cons of using this alternative energy source. Numerous advantages include direct access to available sunlight, constant technological improvement to economically tap solar power, easy installation with reasonable expertise and above all significant reduction in energy costs. However, there are certain obvious constraints too like limited availability of sunlight (not available at night), impact of weather and location on accessing sunlight, rather expensive technology, not widely available technological innovations to store solar energy, need of ample of space to install panels.

Global Competition in Solar Energy

The global competition in this alternative energy sector is severe as solar companies in the world are growing at a relatively fast pace. The initial cost of installation is high but it brings high dividends in the long-term scenario. Solar power plants are emerging as an integral part of the power grid from which the power companies provide electricity.

It is interesting to note that 22,000 companies are in the fray across the world to grab their due share but two Chinese companies, Yingli and Trina, together now account for around 15 per cent of the world solar module market. The scene is completely dominated by China, with nearly half of the world marketshare and unrelenting upward journey.

At the beginning of the century, China's share was less than 1 per cent which has now dramatically jumped to 50 per cent, thanks to the immense government support and incentives which Chinese companies have been enjoying so far. Let us first cast a glance at the Top 10 global players in terms of their solar power production capacity in 2011:

1. LDK Solar (China)

The largest solar power company in the world making PV cells and modules is China's LDK Solar, which burst upon the scene only in 2010. It was a Chinese solar power company and operates out of Xinyu City, Jiangxi, China. The solar panels from LDK Solar use mono-crystalline and poly-crystalline technologies in their solar power plants. An Indian court in July 2018 issued a nearly US$16 million arbitral award to the

Chinese PV company LDK Solar Hi-Tech following its dispute with an Indian renewable energy company, Hindustan Clean Energy (formerly Moser Bear Clean Energy) over a solar panel supply agreement.

2. Sharp Solar (Japan)

Sharp Solar has been operating from Osaka Japan since 1963 and is the oldest solar company but still going strong. With a net power output of 2.8 gw, it operates across Japan, UK, and USA.

3. Suntech Power (China)

This Chinese solar energy company occupies the third place overall in the solar power market. It provides solar power systems in Japan, Germany and the US, apart from China. It has been using mono c-Si, poly c-Si and thin film technologies in its solar panels. It produces an estimated 2.4 gw of solar energy globally.

4. First Solar (USA)

First Solar power company has its headquarters at Arizona, USA and operates in the US, Germany and Malaysia. The company has been in the power production business since 2005, using thin silicon film wafers for its solar cells. The estimated production capacity of the solar company is 2.3 gw.

5. JA Solar (China)

JA Solar is yet another solar power company from the Chinese homeland in operation since 2006. It has an estimated production capacity of 2.2 gw.

6. Canadian Solar (Ontario, Canada)

One of the top solar companies in the world, Canadian Solar has its presence in the Chinese mainland too! With an estimated production capacity of 2 gw solar power, it uses mono c-Si and poly c-Si technologies. Canadian Solar is a relatively new entrant yet has made its way to the sixth place.

7. Trina Solar (China)

Trina Solar is yet another key global player in the field of solar power production. In business since 2004, this solar company had an overall capacity of 1.9 gw as on 2011.

8. Yingli Green Energy (China)

With an estimated production capacity of 1.7 gw, Yingli Green Energy is also from China and has been operating since 2002. It also uses mono and poly c-Si technology in its panels. Yingli is on the eighth position in the Top 10 solar companies list.

9. Hanwha Solar One (China)

Based in China, Hanwha Solar One is involved in both mono and poly-crystalline silicon technology for the production of photovoltaic cells and modules, and has an overall production capacity of 1.5 gw. This company has been providing solar energy services since 2004.

10. Jinko Solar (China)

Jinko Solar has been in the solar power business since 2009. It had an estimated power production capacity of 1.5 gw even in 2011. It uses mono-crystalline and poly-crystalline silicon solar cells in their panels.

The above list conveys much about Chinese dominance. Seven out of ten companies are from China alone. It is noteworthy that these companies sprang up only after 2000. Even the company on the apex slot (LDK Solar) emerged only in 2010. Their pace of growth is a matter of both guess and surprise!

Spectacular Growth of China Solar Energy

The National Development and Reform Commission (NDRC) of China proposed certain plans to increase the construction of solar power plants in the country. The success story of solar power in China can largely be attributed to reduction in production costs as well as massive state backing. In 2010, the aggregate capacity of China's solar power plants was less than 1 gw. Ground-breaking improvements commenced in 2011 when it raised the total volume of solar energy to 3.3 gw. According to one estimate of the European Photovoltaic Industry Association (EPIA), in 2012 China already had PV capacity of 8.3 gw; in 2013 it was 19.3 gw. Even the most pessimistic EPIA scenario has solar power generating 47 gw for China in 2017, while a more optimistic outlook suggests the figure will be 66 gw. A distinct characteristic of solar energy development in China is the new programme of installing solar panels on rooftops (up to 8 gw in 2014), which reduces the costs of grid modernisation in the remote regions of Gansu, Xinjiang and Qinghai. Solar deployment increased almost fourfold in 2013, to an unprecedented 12.1 gw, besting its earlier record of 3.2 gw in 2012. It attained 19.1 gw capacity by the end of 2013 and 50 gw by 2020. China invested a total of $55.2 billion in the clean energy sector, out of which solar received $22.6 billion in 2013.

New laws in China create a favourable environment to support renewable energy projects. These include authorisation of favourable feed-in tariffs for electricity generated by renewable power generation (including solar alternative energy, wind renewable energy, etc.), financial incentives for setting up alternative energy projects, and loans and tax breaks to support various renewable energy projects.

China energy solar will continue to be a major exporter of solar panels that use PV technology and the western world seems to be unable to compete with cheap labour costs, favourable government policies, and easy investment in China. These policies and other aggressive actions are expected to give a boost to China's solar energy industry, which has lagged behind China's wind, nuclear and hydroelectric power sectors.

Chinese Companies Dominating in US, Germany and Japan

In 2013, China, US, Germany and Japan accounted for two-thirds of worldwide PV demand. China was the largest solar market in the world in 2013. Chinese manufacturer Yingli Green Energy was the globe's foremost supplier of solar modules for the second year in a row.

Hubei (China) based Yingli, was a dazzling, high-output producer in three of those four markets of China, US, Germany and Japan. It shipped a whopping 3 gw of solar modules in 2013. Overall in 2013, Yingli occupied the No. 1 slot in both China and Germany, second in the US and ninth in Japan. The Chinese maker expanded its share of the market to 8.3 per cent, up from 7.4 per cent in 2012 when it was already the world's brightest solar player.

In China, Yingli's solar module shipments amounted to 625.3 mw from the first to the third quarter in 2013, leaving behind other Chinese rivals like Trina Solar and Jinko Solar. Yingli dominated the scene in Germany, dwarfing native German companies in their home turf and came out as a clear winner as well as leader, shipping an estimated 583.9 mw of solar modules. Yingli thus became more than double the size of closest competitor Trina. Despite an anti-dumping trade conflict, Chinese solar suppliers continued to maintain significant activity in the German market, with five out of that country's Top 10 suppliers based out of China. Only two suppliers in that elite circle—SolarWorld and Conergy—were German.

For the United States, the No. 3 PV market in the world, Yingli came in second after Arizona-based First Solar. The US PV market grew by more than 50 per cent in 2013, and Yingli's total shipments of 479.8 mw represented just 70 mw short of First Solar's delivery, thus expending strong competition to technologically superior US.

The only market where Yingli did not find a place within the Top 3 was Japan, the world's second-largest solar space, where Yingli ranked ninth after home-grown producers like Sharp and Kyocera. Foreign suppliers do not have an easy time entering the Japanese solar space, even though entry barriers there are not as high as in China. After turning around in the first half of 2013 from a decline in 2012, solar markets worldwide continued their recovery as 2013 came to a close. Driven by strong demand in Asia, global PV installations rose to 9.2 gw in the third quarter, up from 8.7 gw in the second quarter of 2013. As a result, global solar module shipments increased to 10.1 gw during the period, an all-time high as shipments exceeded 10 gw in a single quarter for the first time ever. Then in the fourth quarter, global PV installations grew to 10.6 gw, and shipments rose as well to 10.3 gw. These figures show how the global PV solar module market became highly competitive, a trend that would continue in the following years.

Japanese Jump

Japan has a long history with solar power. Japanese electronics firm Sharp has been developing solar technologies for more than fifty years, and small solar systems have been deployed throughout the country for the past quarter-century. Japan's ongoing efforts to displace nuclear energy and diversify its power sector since the Fukushima nuclear fallout have propelled the nation's clean energy sector forward in recent years. In 2013, Japan became the fastest-growing clean energy market in the world, growing by 80 per cent. Overall, it rose two spots within the G-20 rankings to become the third-leading destination for clean energy investment, attracting $28.6 billion. Investment in Japan's solar sector nearly doubled; it received $28 billion in 2013, almost 30 per cent of the G-20 total. Most of the solar investment was for small-distributed capacity, which has been stimulated by generous feed-in tariff incentives. Japan added 6.7 gw of solar in 2013. By 2020, Japan's installed PV capacity is said to be about 60 gw.

US' Renewed Focus on CSP

The solar energy industry has been one of the fastest growing industries in the United States over the last five years. By the end of 2013, the United States had more than 13 gw of installed solar capacity—nearly 15 times the amount installed in 2008 and enough to power more than 2 million average American homes. Across the nation, solar energy is taking off, with more Americans 'going solar' every day. And, it's not just solar panels popping up on the rooftops of homes; Americans are starting to adopt other forms of solar energy as well.

Concentrating Solar Power (CSP) is a technology that harnesses the sun's energy potential and has the capacity to provide hundreds of thousands of customers with reliable renewable energy—even when the sun isn't shining. This is particularly well suited for its sun-drenched southwestern states. The year 2014 marked a significant milestone in the history of American solar energy. Through sustained, long-term investments by the US Department of Energy (DOE) and committed industry partners, some of the most innovative CSP plants in the world connected to the United States electricity grid in 2013, and five plants of this kind are expected to be fully operational by the end of 2014. One of them is the largest CSP plant in the world; another represents a first-of-its-kind energy storage technology at commercial scale in the United States. Collectively, these five CSP plants will nearly quadruple the pre-existing capacity in the United States.

What is so unique about CSP technology? When deployed with thermal energy storage, it can produce electricity on demand—providing a dispatchable source of renewable energy. Therefore, it can provide electricity whenever needed by an electric utility to meet consumer demand, performing like a traditional base-load power plant. Another distinction of CSP is its ability to integrate with fossil-based generation sources in 'hybrid' configurations. Hybrid systems combine traditional fossil-fuel-powered plants with emissions-free CSP technology to improve the performance of both systems.

Three cutting-edge key technologies being deployed are:

- Parabolic Trough Systems
- Power Tower
- Dish Engine System

Let's have a brief overview of these technologies.

Parabolic Trough Systems

Trough systems use large, U-shaped (parabolic) reflectors (focusing mirrors) that have oil-filled pipes running along their centre, or focal point. The mirrored reflectors are tilted toward the sun, and focus sunlight on the pipes to heat the oil inside to as much as 750°F. The hot oil is then used to boil water, which makes steam to run conventional steam turbines and generators.

Power Tower Systems

Power tower systems also called central receivers, use many large, flat heliostats (mirrors) to track the sun and focus its rays onto a receiver. The receiver sits on top of a tall tower in which concentrated sunlight heats a fluid, such as molten salt, as hot as 1,050°F. The hot fluid can be used immediately to make steam for electricity generation or stored for later use. Molten salt retains heat efficiently, so it can be stored for days before being converted into electricity. That means electricity can be produced during periods of peak need on cloudy days or even several hours after sunset.

Dish Engine Systems

Dish/engine systems use mirrored dishes (about 10 times larger than a backyard satellite dish) to focus and concentrate sunlight onto a receiver. The receiver is mounted at the focal point of the dish. To capture the maximum amount of solar energy, the dish assembly tracks the sun across the sky. The receiver is integrated into a high efficiency 'external' combustion engine. The engine has thin tubes containing hydrogen or helium gas that run along the outside of the engine's four piston cylinders and open into the cylinders. As concentrated sunlight falls on the receiver, it heats the gas in the tubes to very high temperatures, which causes hot gas to expand inside the cylinders. The expanding gas drives the pistons. The pistons turn a crankshaft, which drives an electric generator. The receiver, engine, and generator comprise a single, integrated assembly mounted at the focus of the mirrored dish.

Five promising CSP projects are scattered in the deserts of Mojave, dunes of Nevada, Arizona and California. They are:

- Crescent Dunes—Tonopah, Nevada
- Genesis—Blythe, California
- Ivanpah—Dry Lake, California
- Mojave—near Barstow, California
- Solana—near Gila Bend, Arizona

Indian Solar Energy

India has ambitious plans for wider solar energy penetration, especially in rural areas. The Solar Mission has set ambitious targets of 100 gw for 2022. The current installed capacity is 32 gw.

Overall, the energy sector in India has a total installed capacity of approximately 1,46,753 mw out of which most is coal-based (54 per cent). Hydro contributes around 25 per cent and only 8 per cent is supported by renewable energy sources. The rest of the shortfall comes from natural gas and nuclear energy. Even though the installed capacity is huge, it is estimated that demand for additional 11 per cent of total energy will lead to power shortages. Therefore, tapping into alternative energy sources is imperative for meeting the shortfall.

Solar energy in India meets the needs of the rural sector since many of village clusters are not served by the traditional electric-grid systems. The simplest use of solar energy is the development of solar lanterns. The lamp is a low-cost lighting solution to specifically meet the needs of rural areas where either there is no grid-based electricity or power supply is erratic. Solar lanterns these days use LED or CFL and can be used indoors as well as outdoors. The advantage of solar lanterns is that they are pollution free and tap free sunlight. They use solar PV modules of 5-6 watts and even allow for charging some models of cellphone.

Many people in urban areas of India are beginning to realise the potential of solar panels. The government of India has established various agencies to lead the renewable energy sector to increase energy supply and meet the demand of decentralised rural areas. The strategy is administered through India's Ministry of Energy, energy development agencies in the various States, and Indian Renewable Energy Development Agency (IREDA). The Solar Energy Centre (SEC) falls under Energy Ministry and undertakes activities related to research and development (R&D), including design, development, testing, standardisation, consultancy, training and information dissemination in the field of solar alternative energy.

Hence the solar energy programme in India has vast potential to not only replace energy supply provided by fossil fuels but to supply energy to energy-deficient rural India. However, market growth and commercial realisation will remain heavily dependent on government policies, market needs, prices of PV panels and development of new technologies.

OTHER PROMISING ALTERNATIVES ENERGY SOURCES

Hydrogen and Fuel Cell: Ultra Clean Energy

Though hydrogen doesn't occur naturally on the Earth yet it's always available in other elements like water (H_2O) which is a combination of hydrogen and oxygen. It is also found in many other organic compounds such as the hydrocarbons that make up many of our fuels, such as gasoline, natural gas, methanol, and propane. Hydrogen can be separated from hydrocarbons through the application of heat, a process known as reforming. Currently, most hydrogen is made this way from natural gas. An electrical current can also be used to separate water into its constituent components of oxygen and hydrogen. This process is known as electrolysis. Some algae and bacteria, using sunlight as their energy source, even give off hydrogen under certain conditions.

Hydrogen is high in energy and produces almost no pollution. NASA has used liquid hydrogen since the 1970s to propel its space shuttles and rockets. A fuel cell basically combines hydrogen and oxygen to produce electricity, heat, and water. Fuel cells could be compared to batteries. Both convert the energy produced by a chemical reaction into usable electric power. However, the fuel cell will produce electricity as long as the fuel hydrogen is supplied, never losing its charge. A fuel cell stack may contain a few to hundreds of individual fuel cells layered together. This 'scalability' makes fuel cells ideal for a wide variety of applications, such as stationary power stations, portable devices and transportation.

Fuel cells are a promising technology for use as a source of heat and electricity for buildings, and as an electrical power source for electric motors propelling vehicles. Fuel cells operate best on pure hydrogen. But fuels like natural gas, methanol, even gasoline can be reformed to produce the hydrogen required for fuel cells. Some fuel cells even can be fueled directly with methanol, without using a reformer. In future, hydrogen could also join electricity as an important energy carrier.

Algae Bio-fuel

Scientists have been experimenting and exploring other lesser known energy sources. Micro-algae is one such promising source which has the potential to quench a sizeable chunk of the world's energy demands. Micro-algae produce much higher yields of fuel-producing biomass than other traditional fuel feedstock and don't compete with food crops. Algae's demonstrated yield is about 2,500 gallons of bio-fuel per acre per year. Moreover, micro-algae can be produced in non-arable lands unsuitable for agriculture. Non-arable land in countries like Brazil, Canada, China and the US could be used to produce enough algal bio-fuel to supplement more than 30 per cent of those countries' fuel consumption. For the waste management sector, algal bio-fuels are of particular interest as they can utilise waste water such as landfill leachate.

Bio-waste

Bio-fuels can be produced from a variety of bio-wastes viz. residues/wastes from agriculture and forestry, industrial wastes, Municipal Solid Waste (MSW) and aquatic biomass. Over recent decades, significant R&D and pilot activities have aimed at enlarging the feedstock base for bio-fuels to cover a range of sustainable sources and to develop processing technologies. However, while advanced bio-fuel technology is being developed, large volumes of advanced bio-fuels are not yet available, and production is still largely restricted to demonstration units.

British Airways has been developing a facility in East London to convert 500,000 tonnes of varied waste into 16 million gallons of jet fuel each year. The plant will deploy new Solena's Plasma Gasification (SPG) technology, which involves a combination of plasma arc gasification and the Fischer-Tropsch process to convert waste into biofuel. Compared to conventional gasification technologies, SPG technology could process up to 50 per cent more waste, and can use a variety of feedstocks such as paper, plastics, tyres, chips and forestry residues.

In Sweden, Germany-based energy provider E.ON plans to develop a 200 mw bio 2G plant, which will serve as industrial reference plant for biomass gasification to produce bio-methane.

A number of research projects and companies are developing innovative processes viz. pyrolysis and thermo-chemical conversion to turn a wide range of bio-waste and biomass such as forestry residues, crop residues, waste paper and organic waste into stable, concentrated bio-oil called bio-crude. Shell has developed the HTU (hydrothermal upgrading) process in which biomass is treated with water at high temperature and pressure (300-350°C and 120-180 bar) (a metric unit of pressure) to produce bio-crude. Both heavy and light crude can be produced by this process. The Canadian company, Ensyn has demonstrated its Rapid Thermal Processing Renewable Fuel Oil, using 65 million litres commercially. Biomass Technology Group (BTG) Netherlands, is also carrying out demonstrations of pyrolysis technology to produce biocrude. Tall oil, a residual product of the pulp and paper industry, is also being used to produce bio-diesel.

Ethanol

Ethanol is ethyl alcohol made from renewable plant-based feedstock (sugarcane, maize, corn). Various studies point out that using ethanol in place of gasoline (petrol) helps to reduce direct CO_2 emissions by 34-59 per cent given today's technology. Peer-reviewed, published research shows corn ethanol reduces greenhouse gas (GHG) emissions by 48-59 per cent when compared directly with gasoline. Ethanol contains 35 per cent oxygen, resulting in more complete fuel combustion and thereby reducing harmful tailpipe pollution. Non-toxic, water soluble and quickly biodegradable, ethanol also displaces the use of toxic gasoline components such as benzene, a known carcinogen.

The Top 6 producers of the world are United States, Brazil, Europe, China, India and Canada as per production figures of 2013. Many more countries are becoming ethanol producers, employing traditional crop feedstock and processes. In addition, new technologies for producing ethanol from agricultural, forestry, and municipal wastes and residues are the focus of major research and development efforts around the world. The cost of producing ethanol remains significantly higher than the cost of producing fuel from petroleum. Various governments have given tax incentives to make ethanol competitive with gasoline in the motor fuel marketplace. Continued progress with both conventional and advanced ethanol production technologies could someday result in ethanol production costs competitive with petroleum fuels. E85, a high-level gasoline blend containing 51-83 per cent ethanol is gradually gaining ground. Recently, the Indian government has shown its intent for intensive use of the fuel, which can be used in flexible fuel vehicles (FFVs), which have an internal combustion engine and can run on E85, gasoline, or any blend of the two.

The future looks promising, with Latin America likely to lead the world in fuel ethanol production. This may be explained with the high yields in sugarcane production and the fact that many of these economies are agriculture based. Several projects in Latin America such as Peru, Colombia or the Central American states have been initiated. We may see large trade flows from South America to North America. Another trade flow may be directed at the Asian/Pacific region and here Japan and possibly South Korea could take top position. Moreover, there is the possibility of developing an export flow from South America to the European Union. Several countries in Latin America enjoy duty-fee access to the European market and they would be in a prime position to act as suppliers

Southern Africa could become another potential supplier to the world market because of relatively high sugarcane yields and some underutilised areas. Several South African countries also enjoy duty-free access to the European Union and therefore, some quantities may go there. Another potential export market for distillers in sub-Saharan Africa could be the Far East. India, Thailand and Australia may emerge as smaller to medium sized exporters with South Korea and Japan on the importing side.

Water Hyacinth

Also known as *Eichhornia Crassipes,* water hyacinth proliferates in stagnant water bodies. It is native to the Amazon Basin and is now commonly seen in most Asian countries. Plants are identified by their broad, thick, glossy, ovate leaves and long, spongy, bulbous stalks. They could rise above the surface of the water as much as one metre in height. Till now, they were considered a menace as they obstruct navigation and reduce oxygen content of the water bodies. Yet they have specific qualities which can be utilised to produce both bio-ethanol and bio-gas to power vehicles and to generate electricity. They do possess a phenomenal growth rate. Apart from this, they do not compete with other food crops for land diversification. They tend to grow well in water bodies even with low nutrient regime and can be easily harvested either manually or mechanically. Some demonstrations have been successful in extracting 10 kg gas out of 1 kg plant mass, vindicating ample potential of this plant.

The quest for alternative energy will intensify in times to come as people become increasingly conscious of the importance of alternative energy. Alternative energy has to be given a fair chance to lessen dependence on energy sources. This can be done by making small changes in our everyday routine. We could avoid use of electric or electronic devices for things which could be accomplished manually. Reusing and recycling ought to become an integral part of our daily routine. The best alternative energy remains 'human muscle power', which can never become obsolete.

Coal/Coke Galore, Clamour for More!

THE INVENTION of the coal-burning steam engine by James Watt in the eighteenth century engendered an almost insatiable demand for coal the world over! It heralded a completely new era in which black coal suddenly became the most sought after natural resource. Coal could well be considered mankind's earliest sources of energy, imparting much needed heat and light. China perhaps discovered it first and was known to have used it more than 3,000 years ago, even before Europe and America.

Solid coal, usually brown or black, a carbon-rich material that most often occurs in stratified sedimentary deposits, is one of the most important primary fossil fuels. Over the centuries, it became an essential element of our lives. Though many other alternative source of energy have been discovered and evolved, yet coal ironically remains indispensable and constitutes umpteen nations' main source of energy; it will continue to play a dominant role for many more decades to come.

Ancient and Modern Usage

References to the early uses of coal are scanty and difficult to trace. However, earlier fuels were primarily wood and charcoal derived from it. Straw and dried dung constituted secondary sources. Aristotle referred to 'bodies which have more of earth than of smoke' and called them 'coal-like substances.' Biblical references to coal are often to charcoal rather than to the rock. Coal was used commercially by the Chinese long before it was utilised in Europe. Coal from the Fushun mine in north-eastern China may have been deployed for copper smelting as early as 100 BC. 'Stones' used as fuel were said to have been produced in China during Han dynasty during 206 BC to 220 AD.

The Romans were familiar with its use even before 400 AD; this is amply attested by the coal cinders found among Roman ruins in England. The first documented proof that coal was mined in Europe was provided by the monk Reinier, who wrote of 'black earth' very similar to charcoal used by metal workers. Many references to coal mining in England, Scotland, and the European continent began to appear in the writings of the thirteenth century. Coal was, however, used only on a limited scale until the early eighteenth century, when Abraham Darby of England and others developed methods of using coke made from coal in blast furnaces and forges.

Up to the American Revolution, most coal used in the American colonies was requisitioned from England or Nova Scotia. War-time exigencies spurred small American coal-mining operations such as those in Virginia on the James river near Richmond. By the early 1830s, mining companies had sprung up along the Ohio, Illinois, and Mississippi rivers and in the Appalachian region. As in European countries, the introduction of the steam locomotive imparted a strong fillip to the American coal industry. Sustained expansion of industrial activity in the US and in Europe further augmented coal use.

In modern times, aside from an energy source, coal is also used as a chemical feedstock from which numerous synthetic compounds such as dye, oil, wax, pharmaceuticals and pesticide are derived. Coal is intensively used in the production of coke, which finds application in various metallurgical processes. Coal is a primary source of energy in the production of electrical power from coal-fired thermal power plants. Moreover, coal gasification and liquefaction produce gaseous and liquid fuels that can be easily transported and conveniently stored in tanks.

Emerging New Technology CTL

Escalating oil prices and rising oil imports are a cause of worry for many countries like India. They could tackle the issue by going for CTL Technology. CTL (converting coal to a liquid fuel) is a process referred to as coal liquefaction which allows coal to be utilised as an alternative to oil. There are two different methods for converting coal into liquid fuel: direct liquefaction and indirect liquefaction.

Direct liquefaction works by dissolving the coal in a solvent at high temperature and pressure whereas indirect liquefaction gasifies the coal to form a 'syngas'. The former process is highly efficient, but the liquid products require further refining to achieve high grade fuel characteristics.

An array of products can be made via these processes:

ultra-clean petroleum and diesel, as well as synthetic waxes, lubricants, chemical feedstock and alternative liquid fuels such as methanol and dimethyl ether (DME).

South Africa has been producing coal-derived fuels since 1955 and has the only commercial coal-to-liquid industry in operation today. Not only are CTL fuels used in cars and other vehicles, South African energy company Sasol's CTL fuels have been approved for utilisation in commercial jets. Currently around 30 per cent of the country's gasoline and diesel needs are produced from indigenous coal. There are a number of CTL projects around the world at various stages of development. Liquid fuel from coal can be delivered from an existing pump at a filling station via existing distribution infrastructure and used, without modification, in the current vehicle fleet.

How CTL can be Popularised

Certain technological advancements are still desired in order to make CTL viable and universally applicable. They may include minimising water used in the process, improving process design to achieve near zero emission, ascertaining the optimum degree of mineral matter removal from coal, achieving optimum operating conditions that result in the most cost and energy-effective CTL process. Reduction in processing cost is highly desirable to facilitate large scale production. Moreover, optimising CTL product quality, maximising fuel yield and increasing process efficiency are other areas that need to be further researched.

Global Coal Production Scenario

The importance of coal could well be gauged by the simple fact that coal provides around 30 per cent of global primary energy needs, generates 41 per cent of the world's electricity and is used in the production of 70 per cent of the world's steel.

There has been a consistent rise in global coal production since 1990. The world's coal production was 4,677 mt in 1990 which reached a record level of 7,831 mt in 2012. It recorded a 2.9 per cent year-on-year growth over the preceding year.

Global coal production continued to be at 7,813 mt in 2018. It is interesting that China's coal production increased substantially by 161.68 per cent since 2000 whereas production fell drastically in the US to balance out the gain and the loss.

It would also be worthwhile to look at top coal-producing states in the world. China comes first, producing 3,550 mt annually (2018). India is second, producing around 770 mt annually in 2018. The others in descending order are: US, Indonesia, Australia, Russia, South Africa, Germany, Poland and Kazakhstan. The graphic below makes explicit the global scene.

It's a matter of grave concern that even after the 2015 Paris Accord there has been no let-up in global coal production. Coal burning is attributed to 0.3 degrees C rise in GAT (Global average temperature) out of 1° total rise since 1800.

Asia accounts for three-fourths of global coal consumption today. More importantly, it accounts for more than three-fourths of coal plants that are either under construction or in the planning stages—a whopping 1,200 of them, according to Urgewald, a German advocacy group that tracks coal development. It's certainly an assault on the Paris Accord.

Indonesia is digging more coal. Vietnam is clearing ground for new coal-fired power plants. Japan, reeling from the 2011 nuclear plant disaster, has resurrected coal. China is now scrambling to find new markets, from Asia to Africa. Chinese companies are now building coal plants in 17 countries, according to Urgewald. Its regional rival Japan is in the game too; nearly 60 per cent of planned coal projects developed by Japanese companies are outside the country, mostly financed by Japanese banks.

India too seems unable to abandon massive coal production as of now. Some 50 gw of additional coal capacity is under construction. Many of these plants are meant to replace older, more polluting ones. But coal would not sunset anytime soon, not until there's a cheap and efficient way to store energy from solar and wind energy. Even today, coal accounts for 58 per cent of India's energy mix.

Graphic 4.1: Top 10 Coal Producing Countries in 2012

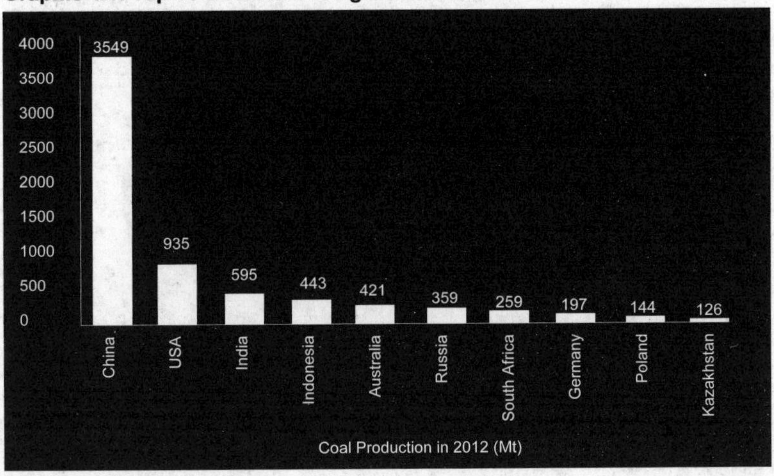

*Data Source: World Coal Association

Global Energy Resources

Coal (hard coal and lignite), crude oil (conventional and non-conventional), natural gas (conventional and non-conventional), uranium and thorium together constitute global energy resources.

Compared to all of the other energy resources, coal continues its overwhelming dominance with a share of around 89 per cent. Way down in the second slot come the resources of natural gas, accounting for nearly 6 per cent, of which the share of non-conventional deposits clearly exceeds the figure for conventional deposits. Crude oil comes in third place, accounting for around 3.6 per cent of total resources. Uranium and thorium account for 1.2 per cent and 0.5 per cent respectively of total global resources. The energy content of all the energy resources put together is a whopping 5,33,526 ej. The graphic on the next page depicts the picture in a lucid manner.

Graphic 4.2: Global Energy Resources and Share of Coal

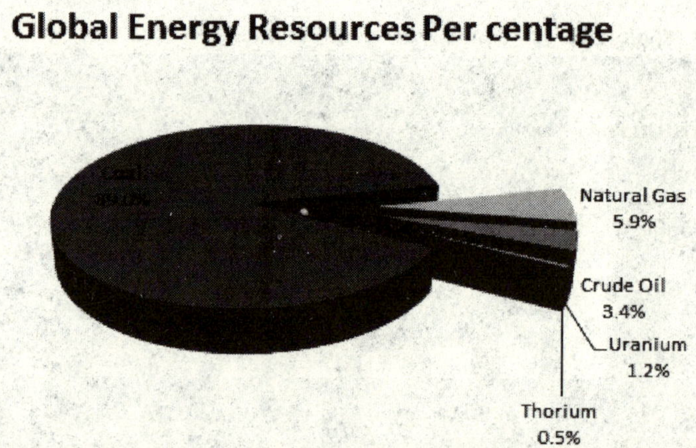

*Data Source: BGR, Germany

India's Coal Resources

India boasts of abundant coal resources but has harnessed little so far. The coal resources of India are available in older Gondwana Formations of peninsular India and in the north-eastern region (Younger Tertiary Formations). Large coal resources are concentrated in Jharkhand, followed by Odisha and Chhattisgarh. Other noteworthy states (with more than 10,000 mt) are West Bengal, Andhra Pradesh, Madhya Pradesh and Maharashtra. It is estimated that India possesses up to 3 lakh mt of total coal resources comprising proved, indicated and inferred resources. A rising pattern is evident since 2009 in coal resources as a result of new explorations and upgradation.

Global Energy Reserves

In terms of reserves, coal is again the undisputed lord commanding the apex position with 56 per cent share of total energy reserves. Crude oil comes next only to coal which accounts for 22.7 per cent of total reserves. Third and fourth slots are occupied by natural gas (18.7 per cent) and uranium (2.7 per cent) respectively. The energy content of all these reserves in 2012 corresponded to 39,910 ej. The graphic below exhibits the global energy reserves with percentage share of each component.

Graphic 4.3: Global Energy Reserves Ending 2012

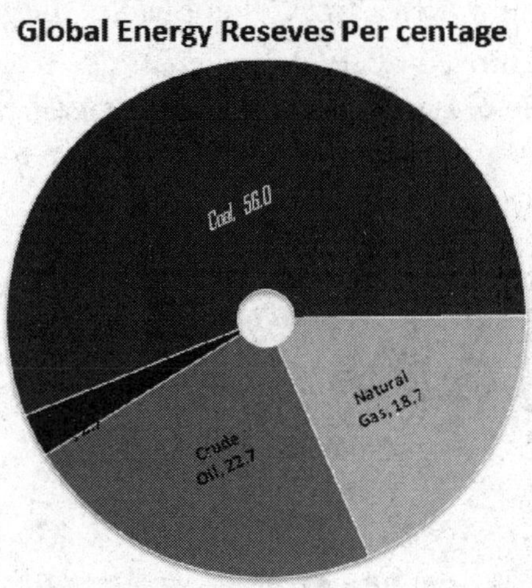

*Data Source: BGR, Germany

Hence it is evident from the global energy resources and reserve analyses that coal rules the roost and occupies the paramount position.

In 2018, coal accounted for 26 per cent of global primary energy consumption, maintaining its position as the second largest energy source after crude oil and first for the electricity generation with 38 per cent of global power generation. Unlike oil and conventional gas deposits, coal fields and their production sites are located in many countries and exploited by many companies.

Countries with Largest Proven Coal Reserves

The US occupies the top position in terms of proven coal reserves, with an estimated reserve of around 237.3 bt at end-2012. Both Russia and China are sitting on huge reserves of around 157 bt and 114.5 bt, respectively. Australia and India are other important countries having reserves of 76.4 bt and 60.6 bt, respectively. The next five are Germany, Ukraine, Kazakhstan, South Africa and Colombia. See the graphic below for insights on proved reserves.

Graphic 4.4: Countries with Largest Coal Reserves Ending 2012

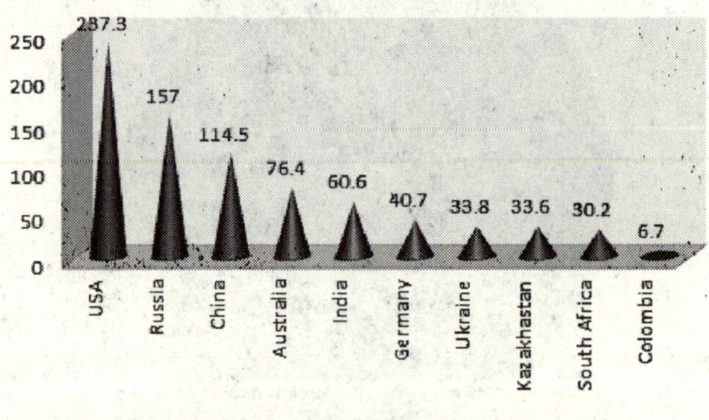

*Data Source: World Coal Association

Prominent Coal Miners and Mines

Let us also look at significant players in the coal mining business. Setting the 2010 production figures as benchmark, the following companies emerge the top players. Coal India, which is one of the seven 'Maharatna' companies of India, reigns supreme. It is expected to make rapid strides in the coming years by means of more overseas acquisitions. Another close competitor is China's Shenhua which is also expanding itself by aggressive foreign acquisitions.

Here is the list of Top 10 with production figures of 2010: Coal India (4,312 mt), Shenhua Group (3,523), Peabody Energy (1,984), Datong Coal (1,505), Arch Coal (1,466), China Coal (1,387), BHP Billiton (1,048), Shanxi Coal (1,019), RWE Power (9,910) and Anglo American (97).

It is interesting to note that the list is dominated by Chinese companies with four—Shenhua, Datong, China Coal and Shanxi Coal—occupying the second, fourth, sixth and eighth positions respectively. It shouldn't come as a surprise seeing the astounding and aggressive expansion they have been embracing since the first decade of this century.

The question of how many coal mines there are worldwide can only be partially answered. Various sources put together arrive at the figure of around 24,000 mines in the entire globe. Underground mining is dominant. A large percentage of mines are small-scale, relatively unproductive mines in China. Production figures from the International Energy Agency (IEA) suggest that only around 60 per cent of total coal production comes from underground mines whereas the remaining 40 per cent comes from opencast mining. The following infographic imparts the full picture of the number of mines reported in

China, the US, India and the UK with break-up of underground and opencast.

Graphic 4.5: Distribution of Coal Mines in World

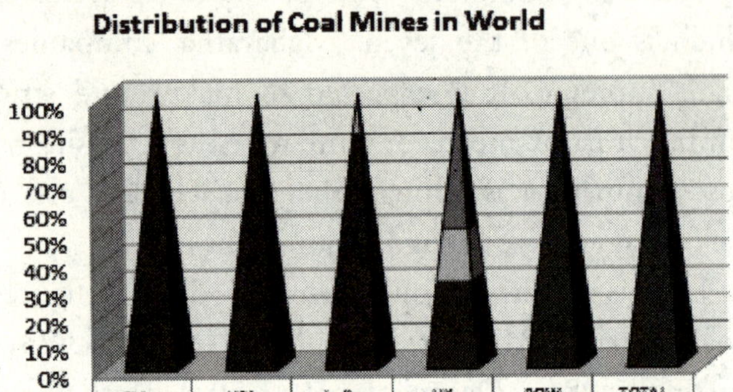

	China	USA	India	UK	ROW	TOTAL WORLD
Opencast %age	5	60	36	72		
Unerground % age	95	40	64	28		
Total Mines	18557	1458	562	46	3377	24000

*ROW: Rest of the World

Source:: International Energy Agency

COAL IMPORTING COUNTRIES

The top hard coal importing countries are Japan, China, South Korea, India, Taiwan, Germany, Turkey, the UK, Italy and Malaysia, together accounting for about 75 per cent of hard coal trade. In 2010, 18 per cent of hard coal imports were made by the EU countries. In Asia, it is China, India, Japan, South Korea and Taiwan who are the traditional buyers and dominate the scene. Together they account for 84 per cent of total coal imports. The surge in China's and India's imports has changed the global coal trade scenario. China became the world's top importer in 2011, taking over the position that Japan occupied for nearly three decades. India occupied the third slot in 2012 surpassing South Korea.

People's Republic of China and India

China now accounts for nearly half of world coal consumption. The figures for China are always dazzling, with trade blossoming by leaps and bounds. In 2011, China alone accounted for nearly half the world's coal consumption. Its consumption reached 3.5 bt equivalent to 1,840 mt oil and is almost three times higher than in 2000. The power sector is by far the largest coal user in China, accounting for more than half of total consumption. Power demand is concomitant to burgeoning economic activity which resulted in the high growth observed over the past decade. The second largest user, the iron and steel industry, also augmented its coal consumption tremendously, driven by the surge in steel production. Chinese crude steel production grew from 490 mt in 2007 to 716.5 mt in 2012. China now accounts for nearly 50 per cent of the entire global crude steel production. Other large users in China include the cement and chemical industries.

India is now the second largest producer of coal (716 mt in 2018) but domestic production is insufficient to cover the country's fast-growing needs, emanating from both electric power producers and steel manufacturers. In 2012, imports were around 134 mt, up 15 per cent over 2011. India imported 208.27 mt in 2017-18 and exported 1.66 mt only in 2017-18. While this figure is much lower than China's imports, recourse to imports is crucial in a country where shortfall of coal and electricity are concurrent. The government embarked on an extensive construction programme of large power plants, most often located on the coast and powered by coal, resulting in high imports.

India has an ambitious electrification programme and in 2006 launched an initiative for the construction of 14 coal-based Ultra Mega Power Projects (UMPPs), each with a capacity of 4,000 mw. Although most plants were delayed, India's power capacity reached 182 gw at end 2011, with 54 per cent based on coal (100 gw). The problem is not capacity addition, but availability of fuel. Though India has plentiful coal reserves, estimated at 315.14 bt in 2017 yet they are of poor quality with high ash content. Moreover, they are mainly confined to eastern and south-central parts of the country, far from centres of consumption. The logistic bottlenecks put regular pressure on domestic supply. CIL produces 80 per cent of domestic coal.

The government regularly introduces new reforms to improve the efficiency of the coal sector, but these have so far not borne fruit, due to the many hurdles to be surmounted. In order to secure domestic supply, government is pursuing a three-pronged strategy: increasing production, establishing a coordinated import policy, and acquiring overseas mines.

The Japan-South Korea-Taiwan Triangle

Japan, South Korea and Taiwan constitute the traditional Asian buyers group. These three countries do not produce coal and solely rely on imported coal to fuel coal-based power generation and to manufacture steel products. With Chinese exports falling, these buying countries have switched to using Australian, Indonesian and US coals in the past few years.

Prominent Importing EU Countries

Coal demand in Europe has been shrinking over the years as a result of coal plant closures and rising renewable source-based power generation, together with a combination of higher carbon and low natural gas prices, as well as slow economic growth and improved energy efficiency.

In 2018, total seaborne coal imports into Europe (excluding Russia and Turkey) totalled 138.0 mt, or about 12 per cent of global seaborne coal trade. This included 42.4 mt discharged in the Netherlands (mainly at Rotterdam, Amsterdam, Ijmuiden, Eemshaven), 18.3 mt in Spain (mainly at Gijon, Tarragona, El Ferrol, Carboneras, La Coruna), 13.2 mt in Italy (at Brindisi, Civitavecchia, La Spezia, Porto Torres, Taranto, Trieste), 13.1 mt in Germany (Bremen, Hamburg, Nordenham, Rostock, Wilhelmshaven), 8.2 mt in Ukraine (Nikolaev, Odessa, Yuzhny), 7.8 mt in France (Dunkirk, Fos), 7.3 mt in Poland (Gdansk, Gdynia, Swinoujscie), 6.7 mt in the UK (Belfast, Immingham, Port Talbot).

The largest importer of coal in Europe in 2019 was the Netherlands with 35.9 million tonnes in the first 11 months of 2019, followed by Germany with 10.8 mt, Spain with 10.5 mt, Italy with 7.9 mt, Ukraine with 6.6 mt, France with 6.6 mt, Poland with 6.1 mt, the UK with 3.3 mt. The respective figures for the first eleven months of 2018 were: the Netherlands with 38.7 mt, Spain with 16.6 mt, Germany with 11.9 mt, Italy with 11.8 mt, Ukraine with 7.6 mt, France with 7.1 mt, Poland with 6.2 mt and the UK with 6.0 mt.

INDIAN COAL BLOCK ALLOCATION MESS

Coal extraction and mining was largely a state owned affair in India. There were no specific criteria for the allocation of coal blocks to PSUs or private entities in India till 1993, when the Ministry of Coal started according coal blocks to private parties for captive mining on recommendations of the Screening

Committee or through direct allocation. The concept of allocation of captive coal blocks through competitive bidding was first announced in 2004.

In a draft report issued in March 2012, the Comptroller and Auditor General of India (CAG) office accused the government of allocating coal blocks in an inefficient manner during the period 2004–2009.

The CAG feels the loss could have been avoided, or at least have been lower, as the procedure for auction of the precious natural resource could easily have been put in place by 2006. The failure to do so meant that 25 firms, including Essar Power, Hindalco, Tata Steel, Tata Power and Jindal Steel and Power, received a windfall gain through coal blocks on nomination basis, instead of competitive bidding.

A report by Standing Committee on Coal and Steel tabled in Parliament on 23 April 2013 stated that all coal blocks distributed between 1993 and 2008 were done in an unauthorised manner and allotment of all mines where production is yet to start should be cancelled. It further recommended that all 'personnel' who have been involved 'directly or indirectly' in the allocation process 'should be investigated for their role'. There was no transparency in the allocation process and the exchequer did not get any revenue from allocation of the blocks. It also pointed out that the allocations between 1993 and 2004 were done without any advertisement or public information.

SPURT IN ASIAN DEMAND AND COAL CRUNCH

Many nurture the notion that coal is an antiquated, brutish relic that would soon be banished from the global energy scene. There is certainly a lot to dislike about coal. The biggest criticism of the coal and coal mining industries is their enormous potential for environmental pollution. But strangely, coal is still responsible for meeting almost half the energy needs of the entire world. It plays a vital role in steel production as well as in the manufacturing of cement. Its banishment is highly unlikely in the coming couples of decades.

Coal deposits take thousands of years to form naturally and are not replaceable at the pace they are consumed. According to the Peak Coal Theory, global coal output is likely to reach a plateau by 2025, albeit after initially rising over the next few years. Thereafter, it is likely to be on a permanent decline, marking the 'beginning of end' of this vital resource.

A number of countries in the European Union, including France, Italy and the United Kingdom are facing critical shortages of natural resources, including coal. A report disseminated a few years back by the Global Sustainability Institute at Anglia Ruskin University revealed that some countries had less than a year's energy resources remaining and are almost entirely dependent on imports. France had less than a year's worth of its own reserves of oil, gas and coal. Italy had less than a year of gas and coal, and only one year of oil. The UK was left with 5.2 years of oil, 4.5 years of coal and three years of gas reserves.

These countries subsequently adopted better strategies for sustaining their resources like reducing domestic extraction and augmenting imports. Generally, these countries are now exploiting resources in poor countries, which may be good for them but not for the planet as a whole.

A cursory glance at the list of top coal-importing countries shows that the Top 5 are China, Japan, India, South Korea and Taipei, all in Asia. Since 2009, China has been the driving force in global coal markets. Coal imports in China exploded during the preceding five years, jumping to a peak of 35 mt per month in December 2012—a 1,500 per cent increase from 2008 levels. Coal imports of China are still holding at a relatively high level. Thermal coal prices skyrocketed at $140 per metric tonne way back in 2008 when Chinese import demand began to dramatically shoot up.

Another demand driver from Asia is South Korea. Rising coal consumption in that country and a negligible production level have caused the country to rely heavily on imports over the past several years. South Korea held only 139 million metric short tonnes (mmst) of recoverable coal reserves in 2010. The country's coal production of 2.3 mmst was a fraction of its 136 mmst consumption in 2012. As a result, South Korea has emerged the fourth-largest importer of coal in the world, following China, Japan, and India. Coal consumption in South Korea increased by 55 per cent between 2005 and 2012, driven primarily by growing demand from the electric power sector. The electric power sector accounts for 62 per cent of the country's coal consumption, while the industrial sector accounts for most of the remaining amount.

Japan was the largest global coal importer for three decades until 2012 when China surpassed Japan by a wide margin. Coal remains an important fuel source in Japan and accounted for nearly 27 per cent of Japan's electric capacity in 2010. Domestic coal production stopped by 2002, and Japan began importing all of its coal, primarily from Australia. Japan imported 194 mmst coal in 2011. Coal has remained a base load fuel for power in Japan, and coal consumption has hovered around 200 mmst since 2004. Two more coal-fired power plants with a combined capacity of 1.6 gw came online in northern Japan in 2013. Also, 2 gw capacity at the Haramachi plant came back online in 2013 after being decommissioned following damage in the 2011 earthquake. Japan is encouraging utilities to build more coal-fired capacity in the aftermath of the earthquake by lessening the environmental regulations for construction and for burning coal. Coal is expected to displace some of the more expensive oil-fired generators used for power.

GLOBAL VYING FOR COAL MINES

The global need for coal is obvious. Demand-production mismatch and non-availability of any particular resource force countries to turn to overseas shores in order to secure safe supply. In recent times, Indian energy companies have rushed headlong to buy coal assets overseas to meet ever-increasing domestic and industrial demand. India is already seeing a steady rise in coal imports, up by 8 per cent in 2011. There has been a profusion of cross-border deals. Indian energy companies began spending in earnest since 2008-09.

State-owned CIL acquired A1 and A2 mining blocks in Tete, Mozambique (Africa). It won a five-year licence for exploration and development of these two coal blocks in August 2009. These blocks are believed to contain a variety of coal, from the low grade to the best variety coking coal. Roughly 20 per cent of the deposits in these blocks are expected to be of coking coal, the remaining are expected to be thermal coal, which can be used as fuel in power plants. CIL has also floated its overseas arm for the African operations namely Coal India Africana. It has made significant progress in identifying several prospective coal resources in select countries. Adani Power, one of the India's largest private sector coal importers, bought assets from Australia's Linc Energy. It followed this in May 2011 with the purchase of the Abbot Point Coal Terminal in Australia for almost $2 bn. In September, GVK, another power group of India, paid $1.26 bn for assets owned by Australia's Hancock Group. Another Indian company Mercator Lines, though a logistic and coal transportation company entered into coal business in 2008. Through its subsidiary Oorja Holding, it acquired overseas coal mines in Indonesia and Mozambique with substantial coal resources. They are all new greenfield projects. Two mines acquired near town Tanah Grogot in East Kalimantan, Indonesia had reserves of 10-12 mt of coal and production started in 2008. Another greenfield in Batuah village, East Kalimantan, has an area of 914 hectare (ha) with estimated coal reserves up to 60 mt. Production started in August 2012. Drilling has been carried out in 10,000 ha area of Mozambique mines in Africa.

In the face of burgeoning global coal competition, Lanco Infratech Ltd (LIL) of India made a strategic move into the global resources industry. The company, through its Australian subsidiary, acquired Griffin Coal in the year 2011. Griffin is considered the largest individual supplier of thermal coal to Western Australia's industrial coal market with over 1.2 bt of coal resources. Griffin currently mines around 4 mt per annum (mtpa), which essentially caters to domestic customers with an export quantity of around 0.75 mtpa through Kwinana port. From the level of production of 4 mtpa, Lanco plans to ramp up the mining production capacity four-fold to around 18 mtpa by the year 2018.

Similarly, other top business houses from Tata Power and Anil Ambani's Reliance Power to London-listed Essar Energy have also been muscling in, striking deals in Australia, Indonesia and Africa.

To meet China's flourishing demand for every natural resource, outbound investment has become a major trend in recent years, and will continue to grow further with no signs of abatement. Chinese companies are still looking for more attractive opportunities overseas and its largest coal enterprise Shenhua Group has adopted an aggressive approach for seeking assets overseas. In December 2013, it won a licence to develop the Zashulanskoye deposit in the Transbaikal territory with Russia's En+ Group. The deposit is close to the Russia-China border. Shenhua intends to invest nearly $910 million in the exploration of coal mines and building infrastructure at the Zashulanskoye site.

Shenhua also signed an agreement with the government of Mongolia to develop the Tavan Tolgoi thermal coal deposits in southern Mongolia and to build rail links from that site to China. The project is widely believed to be home to the world's largest undeveloped coal deposit. The development of the project was postponed several times due to the political situation in Mongolia. On October 2013, Shenhua and three Mongolian partners met in Beijing to sign an MoU regarding the building of railroads and the coal trade. Mongolia contains high quality coking coal that China lacks.

There were also reports in June 2013 that Shenhua intended to collaborate with partner companies to bid for Rio Tinto's Australian coal mine assets, valued at around US$3.2 billion. Shenhua has its eyes on 'more than one' Australian project at present. China is turning its attentions toward cleaning up the environment, which means that the nation is shifting toward cleaner fuels. Australian coal is exactly the kind of high-quality coal that China craves for.

Some changes were effected for overseas deals with effect from May 2014. Deals below $1 billion do not need any government approval hence Chinese overseas acquisitions are set to expand further.

In March 2007, Korean company CircleOne became the second Korean company to gain resource development rights from the KBB in Indonesia, the largest overseas mine that any Korean company has gained access to. KBB is on Sumatra Island, Indonesia, and covers 18,440 ha, about one-third the size of Seoul. Several Korean companies had been vying for the project, including SK Networks, which teamed up with CircleOne to jointly dig up the coal there. Peabody Energy, the largest US coal producer by output a few years back, entered into a set of deals that would expand its access to booming markets in the Asia-Pacific. In July 2011, the company made a joint bid with ArcelorMittal to acquire Australian coal miner Macarthur Coal for $US5.05 billion.

ROAD AHEAD AMIDST SOARING GLOBAL POLLUTION

Global interest in coal is not going to diminish in the near future. However, the US and Canada may register decline in coal use due to massive shale gas extraction. Though much hue and cry is being made against its profuse use and demand for concrete action against soaring air pollution is vociferously raised at global levels from time to time, concrete action remains elusive. Since one tonne of coal burnt produces nearly

2.4 tonnes of CO_2, reduction of this high volume is paramount and of utmost importance. Data recorded at Mauna Loa Observatory point out that atmospheric CO_2 has decisively breached critical 400 parts per million (ppm) mark and even managed to stay above averaging 401.88 ppm in May 2014. As per NOAA State of Climate Report 2018, global CO_2 was 407 ppm in 2018. The earth capacity of carrying 350 ppm was already surpassed way back in 1988.

This is certainly a grim situation and calls for a judicious and prudent use of coal.

The only way forward seems adoption of Clean Coal Technologies (CCT) on a wider scale. CCTs are not new but several generations of technological advances have led to more efficient combustion of coal with reduced emissions of sulphur dioxide and nitrogen oxide. They certainly facilitate the use of coal in an environmentally satisfactory and economically viable way. Researches in the field have led to numerous lower cost, more efficient and environmentally compatible technologies for electric utilities, steel mills, cement plants and other industries. Currently available CCTs include Pressurised Fluidised-bed Combustion (PFBC), Integrated Gasification Combined Cycle (IGCC), Flue Gas Desulfurisation (FGD), Selective Catalytic Reduction (SCR), Electrostatic Precipitators (ESPs), Low Nitrogen Oxide Burners and Combined heat and power (CHP).

Carbon Capture and Storage (CCS) is another evolving technology under CCT. CCS aims at capturing and storing CO_2 emissions in geologic formations like abandoned gaseous deposits, saline aquifers or deep oceans where it dissolves under pressure. New technologies are also under development subsuming CCS which include post-combustion capture

(PCC), Pre-combustion capture and Oxy-Coal Combustion (OCC). Post-combustion capture reduces flue gases using an amine solvent and chilled ammonia whereas Pre-combustion capture deploys IGCC to isolate and capture CO_2 before it is released. Oxy-Coal combustion uses pure oxygen in the boiler to significantly reduce the dilution of CO_2 in the exhaust gas stream. CCS has demonstrated its utility and has successfully been deployed in several countries. Further R&D is warranted to improve the efficiency and efficacy of the technology.

Efficiency improvements are integral to CCTs, desired to raise plant efficiency and reduce carbon dioxide and other emissions. While some efficiency technologies are commercially available as of now, others such as Ultra Supercritical Pulverised Coal (USPC) and IGCC require continued research, development and demonstration. Other technologies such as coal liquefaction and gasification could well be pursued to produce low cost, secure alternatives to oil and natural gas for use in electricity generation and transportation.

Coliseums of Water Wars

As I travel around the world, people think the only place where there is potential conflict over water is the Middle East, but they are completely wrong. We have the problem all over the world

—Kofi Annan, Former Secretary General, the United Nations

I think the risk of conflicts over water is growing–not shrinking–because of increased competition, because of bad management and, ultimately, because of the impacts of climate change

—Peter Gleick, President Pacific Institute

THINKERS LIKE Kofi Annan and Peter Gleick state unequivocally that water resources worldwide are under increasing pressure due to a combination of factors. In some regions of the world, acute stress could be magnified or become chronic. In academic, political and media circles, there is concern that this will inevitably lead to violent conflicts.

Conflicts over water have been bubbling up in many parts of the world, from China and India to the Middle East and southern Africa. As the world's population grows, so does its thirst. Not every nation can quench it equally. The looming water crisis threatens political stability and economic development in a number of developing and emerging economies, with dire implications for global peace and harmony.

Nor are water wars a recent phenomenon. There is historical evidence that the world's earliest water war took place some 4,500 years ago in Sumerian civilisation (present-day Iraq) when the armies of Lagash (a fertile town nestled between the Tigris and the Euphrates) and Umma entered into a fierce conflict over the issue of water control. Umma cut off water supply to the city of Lagash, hampering domestic produce and trade via waterways, effectively crippling commerce and sending prices on, all commodities spiralling. 'Enannatum, ruler of Lagash, went into battle,' reads an account carved into an ancient stone cylinder, leaving 60 soldiers dead on the bank of the canal.

Though large-scale military combat has not been witnessed in modern times, water wars are more severe and fiercer than ever before, often bringing multitudes out on the streets voicing their anguish. As the stakes become higher as a result of Climate Change and vaulting population, the planet may witness a marked change in the way the nations are prepared to defend or acquire their rights over freshwater resources.

Billions of people even today do not have enough access to safe water. The UN projects that half the nations of the world would face water shortage or anxiety by 2025. It further conjectures that by 2050, water scarcity could affect three-quarters of the world's population. Between 1900 and 1995, global water consumption is said to have risen six-fold—more than double the population growth.

The Food and Agriculture Organisation (FAO) calculated in a 2007 report, 'Coping with Water Scarcity—Challenge of the Twenty-First Century' that if all the freshwater were divided equally among the global population, there would be 5,000-6,000 cum of water available for everyone every year. The internationally accepted definition of scarcity is when people have water availability less than 1,700 cum per person which includes water used for drinking and the production of food. These figures indicate that there is an abundance of water and certainly no global shortage. As the FAO report stated, however, it is not a question of total world supply but rather of the distribution of that water. Water is distributed unevenly across the globe, so is the world's population. It then follows that countries with the acutest water scarcity issues have high population density in addition to relatively low availability of freshwater. Some countries, by virtue of their location and topography, have access to the sources of freshwater; others are dependent on the goodwill of their neighbours to ensure that the waterways, basins and rivers flow unhindered past their boundaries.

There are 263 international basins that cross the political boundaries of two or more countries. These basins cover approximately half of the earth's surface areas and account for an estimated three-fifths of the global freshwater flow and are inhabited by 40 per cent of the world's population.

Historically, despite the complexity of many water disputes, the majority have been handled diplomatically rather than through military means. In the past half century, 37 acute disputes have been reported involving violence while 150 treaties have been signed during this period. According to the FAO, more than 3,600 treaties related to international water resources have been drawn up since 805 AD.

China, Cambodia, Laos, Thailand and Vietnam have managed to cooperate within the Mekong River Commission (MRC) since 1957, even in the face of political rivalry. Similarly, Israel and Jordan have held regular deliberation over sharing the Jordan river, while the Indus River Commission between India and Pakistan continues to function. In 1999, a framework involving ten countries was agreed upon for the Nile river basin.

A detailed scrutiny reveals Jordan topping the list with 29 negative or conflictive events involving its neighbours whereas the Danube enjoys high cooperation between upstream and downstream states. A study conducted for the period 1948 to 1999 filters out around 1,800 events affecting 124 countries. Events included range from a formal declaration of war, political or military hostile actions to strategic alliances or treaties. Out of these 1,831 events, 28 per cent were conflictive, 67 per cent cooperative and the remaining 5 per cent were neutral or insignificant. The majority of events were associated with basins in North Africa and the Middle East, sub-Saharan Africa and Eastern Europe, followed by South-east and South Asia and South America. The Middle East and the North African region show the lowest level of cooperation while West Europe exhibits the highest.

International Laws/ Framework for Water Sharing and Use

Sincere efforts have been made in the past two decades or so to avoid conflicts over trans-national rivers by putting in place appropriate guidelines and framework on water sharing and usage. However, they either don't seem to work or are side-tracked and disputes never subside, particularly when the conflicting states are on unequal footing. Let's take a cursory glance at existing laws and provisions.

United Nations conventions, bilateral or trilateral water treaties between the stakeholder nations, generally constitute accepted norms but many countries brazenly flout sacred clauses enshrined into them. The United Nations has also worked in different regions to bring about cooperation through River Commissions and other frameworks.

UN Convention on Non-Navigational Uses of International Watercourses 1997

The convention calls for reasonable and equitable use of shared watercourses, enjoins countries not to do appreciable harm to others, proclaims environmental protection standards and emphasises the sharing of water-related information. It also warrants considering and establishing linkages between surface water and groundwater. However, its ambiguity leaves the interpretation of these issues up to individual countries. The convention, crucial in establishing international law governing water, has been ratified by 36 countries so far, including Finland, Germany, Hungary, Iraq, Jordan, Lebanon, Libya, Namibia, Netherlands, Norway, Portugal, Qatar, South Africa, Sweden, Syria and Uzbekistan.

UN Agenda 21

In addition to the above convention, in 1992 the United Nations adopted the non-binding Agenda 21, which relates to the protection of the quality and supply of freshwater resources.

UN Law of Trans-boundary Aquifers: An aquifer is a layer formation of rock or soil able to hold or transmit water. An aquifer may contain either fresh water or fossil water. In 2008, the UN General Assembly adopted the Law of Trans-boundary Aquifers resolution, which encourages nations to make bilateral or regional arrangement for the sustainable management of their trans-boundary aquifers.

REGIONS PRONE TO HIGH RISK OF WATER WARS

Basin areas that show signs of present or future conflict may invariably have high density of population, low per capita GDP, hostility with neighbouring states/basins and politically charged minority groups. Apart from these, proposed large dams or other water development projects and absence of any freshwater treaties are often abetting factors.

The following rivers have been identified as high risk for potential unrest:

Asia: The Ca (Laos, Vietnam), Fenney (India, Bangladesh), Ganges-Brahmaputra-Meghna (India, Bangladesh, Bhutan, Nepal, Burma, China), Han (North and South Korea), Indus (India, Pakistan, China, Afghanistan), Irrawaddy (India, Myanmar, China), Karnaphuli (Bangladesh, India), Red (China, Laos, Vietnam), Saigon (Cambodia, Vietnam), Song Vam Co Dong (Cambodia, Vietnam) and Yalu (China and North Korea).

Africa: Cameroon and Nigeria

Balkans: Drin (Albania, Macedonia, Serbia, Montenegro).

China's Dams: Aggravating Regional Discord

China has proclaimed its supremacy in massive dam-building

technology after successful completion of its monumental opus in the form of the Three Gorges Dam (1994-2011) on Yangtze river. The Three Gorges Dam project was indeed a herculean and towering task to accomplish with numerous engineering, topographical, environmental, socio-economic constraints and challenges. Chinese overcame virtually all these challenges and the impossible was finally rendered possible. Chinese prowess in dam building now draws considerable awe and respect both sides of the Atlantic, across continents.

Chinese hydroelectric and state-owned enterprises have embarked upon a massive dam-building spree in all the conceivable geographies. Before 2000, it had about eighteen dam projects overseas and now it boasts of around 350 such projects worldwide,. a mind-boggling number indeed!

Several other competing nations pale in comparison with China. Many countries now prefer Chinese companies if the project is visualised to draw controversy. Chinese companies bring with them massive fleets of workers often supported by Chinese military in case of imminent fray or possible clash. Chinese companies remain taciturn and seldom disclose vital information pertaining to ongoing/proposed projects. They deal with public unrest and demonstrations with a heavy hand, often ignoring defined rules of environment, rehabilitation and resettlement.

These attributes seem to go in favour of Chinese companies especially when the project is in disputed territory or has the potential to arouse significant controversy. Be that as it may, they have been generously awarded many dam building projects everywhere from Asia to Latin America. Data collated from secondary or tertiary sources (reputed NGOs, watch groups, civil society groups) point to their dam building ventures in at least 74 countries. Out of 367 projects, South Asia alone has 118 dam projects. They command over 45 per cent (of total overseas projects) in the Asian continent itself. Their impressive winning streak does not end in Asia alone but goes far beyond that: Europe (12), Middle East (3), Latin America (24), Africa (85) and Pacific (3).

Though the practice of building large dams has received a considerable jolt in the West, it thrives in other parts of the globe, particularly Asia and Africa. China has completed around 75 overseas dam projects by now; equal numbers are under construction and it is further poised to take up approximately 130 small, medium and large dam projects spread all over the world. Such massive exercises are going to fuel discord between the stakeholder states and also among riparian local populace. The water situation is already fluid and highly sensitive seen in conjunction with growing scarcity of fresh water resources, declining rainfall, Climate Change, augmented water use and water demand.

FLASHPOINTS IN VARIOUS GEOGRAPHIES

Though the building of mega dams has gradually diminished in West Europe and the US, it continues to thrive in Africa, Asia and Latin America. Every large dam now attracts resistance and is surrounded by a number of controversies. The legacies and controversies of the world's 45,000 already constructed large dams continue to cause conflict. Few rivers now remain virgin, untouched by some sort of dam/barrage. Dislocated populations, estimated at 40-80 million, have frequently been rehabilitated with minimal or no compensation altogether. In most instances, they are resettled in marginal lands, and in the majority of cases have either become poorer or remained impoverished. Large-scale alteration of natural hydrologic regimes has had massive impacts on fisheries, water-based livelihoods, aquatic ecosystems and environmental services as a whole. Large reservoirs emit huge amounts of greenhouse gases, aggregately, as much as 4 per cent of all human-induced GHG emissions. Indeed, the number of river-dependent people potentially affected by dam-induced changes in river flows and other ecosystem conditions remains quite high.

Here is a description of the global flashpoints or the dams/projects which have hogged considerable limelight in the past or still remain a raging issue crossing the continents.

The Grand Ethiopian Renaissance Dam

This conflict between Egypt and Ethiopia has come to such a pass that Egypt is now considering raising the dispute in the International Court of Justice (ICJ) Hague after failure of successive rounds of talks with the Ethiopian government.

In order to grasp the issue of discord, let us trace back the history of the Grand Ethiopian Renaissance Dam, formerly known as the Millennium Dam and sometimes referred to as Hidase Dam. It is currently under construction on Blue Nile River, situated about 40 km east of Sudan in the Benishangul-Gumuz region of Ethiopia with intended potential of 6,000 mw. It would be the largest hydroelectric power plant in Africa when completed, as well as the thirteenth or fourteenth largest in the world, sharing the spot with Krasnoyarskaya. The full reservoir at 63 bcum will be one of the continent's largest dams. The potential socio-economic and environmental impacts of the dam have been the source of regional controversy. The government of Egypt, a country which relies heavily on the waters of the Nile, protests against the dam. Its political leaders have even discussed ways to sabotage the project.

The eventual site for the Grand Ethiopian Renaissance Dam was identified by the United States Bureau of Reclamation (USBR) during a Blue Nile survey conducted between 1956 and 1964. The Ethiopian Government surveyed the site in October 2009 and August 2010. In November 2010, a design was submitted. On 31 March 2011, a day after the project was made public, a $4.8 billion contract was awarded without competitive bidding to Salini Costruttori and the dam's foundation stone was laid on 2 April 2011 by Ethiopia's Prime Minister Meles Zenawi. A rock-crushing plant has been constructed along with a small air strip for fast transportation.

Egypt, which lies downstream, opposes the dam which it believes will reduce the amount of water that it gets from the Nile. Zenawi argues, based on an unnamed study, that the dam would not reduce water availability downstream and would also regulate water for irrigation. In May 2011, it was announced that Ethiopia would share blueprints for the dam with Egypt so the downstream impact could be examined.

The dam was originally called 'Project X'. After its contract was announced, it was called the Millennium Dam. On 15 April 2011, the Council of Ministers renamed it Grand Ethiopian Renaissance Dam. In March 2012, the Ethiopian Government announced an upgrade to the power plant design, increasing it from 5,250 mw to 6,000 mw. Ethiopia has potential for around 45,000 mw of hydropower. The dam is being funded by government bonds and private donations. The dam is slated for opening in by 2022.

No environmental and social impact assessment for the dam has been published so far. It appears that at least some form of environmental study has been undertaken as the press reported that an international panel reviewed an environmental study in 2012. The NGO International Rivers has commissioned a local researcher to make a field visit because so little information is available.

Public consultation about dams in Ethiopia is affected by the political climate in the country. International Rivers reports that conversations with civil society groups in Ethiopia indicate that questioning the government's energy sector plans is highly risky, and there are legitimate fears of government persecution. Because of this political climate, no groups are actively raising issues surrounding hydro-power dams, nor publicly expressing concerns about the risks. In this situation, extremely limited and inadequate public consultation has been organised during the implementation of major dams. In June 2011, Ethiopian journalist Reeyot Alemu was imprisoned after she raised questions about the proposed dam. The staff of International Rivers received death threats. In the meantime, the leadership called opponents of the project 'hydropower extremists...bordering on the criminal' at a conference of the International Hydropower Association (IHA) in Addis Ababa in April 2011. At the conference, the Ethiopian state power utility was embraced as a 'Sustainability Partner' by the IHA.

River Blue Nile

The Blue Nile is an important tributary of international river Nile. The Blue Nile originates from Lake Tana in the Ethiopian Highlands and flows about 1,400 km to Khartoum, where it joins the White Nile to form the Nile. As much as 90 per cent of the water and 96 per cent of the transported sediment carried by the Nile originates in Ethiopia, with 59 per cent of water from the Blue Nile. The erosion and transportation of silt only occurs during the Ethiopian rainy season in the summer, when rainfall is especially high on the Ethiopian Plateau; the rest of the year, the great rivers draining Ethiopia into the Nile (Sobat, Blue Nile, Tekeze and Atbarah) have a weaker flow. The flow of the Blue Nile varies considerably over its yearly cycle and is the main contribution to the Nile flow's large natural variation.

River Nile Water Sharing Dispute

The Nile is an 'international' river as its water resources are shared by eleven countries, namely, Tanzania, Uganda, Rwanda, Burundi, DRC (Democratic Republic of the Congo), Kenya, Ethiopia, Eritrea, South Sudan, Sudan and Egypt. The Nile is the primary water resource and life artery for Egypt and Sudan.

The Nile's water has affected the politics of East Africa and the Horn of Africa for many decades. Uganda, Sudan, Ethiopia and Kenya have complained about Egyptian domination of its water resources. The Nile Basin Initiative promotes a peaceful cooperation among those states.

Though several attempts have been made to establish agreements between the countries sharing the Nile waters, it is very difficult to have all these countries agree with each other, given the self-interest of each country and their political, strategic, and social differences. On 14 May 2010, Ethiopia, Rwanda, Tanzania and Uganda signed a new agreement at Entebbe, Uganda on sharing the Nile water even though this agreement raised strong opposition from Egypt and Sudan.

Ideally, such international agreements should promote equitable and efficient usage of the Nile basin's water resources. Without a better understanding about the availability of the future water resources of the Nile, we could expect more conflicts between these countries relying on the Nile for water supply as well as economic and social developments.

Turkish GAP project: Listless Syria, Distressed Iraq

Guneydogu Anadolu Projesi (GAP) is a mammoth dam-building scheme of Turkey that envisages the construction of 22 dams and 19 hydropower plants across the Tigris-Euphrates basin. The scheme requires hundreds of kilometres of irrigation canals and is expected to cover 75,000 sq km, around 10 per cent of the surface area of Turkey. As Turkey increases its water use, there has been significant reduction in the flow of water in downstream areas of both the Tigris and Euphrates rivers. It is largely believed that, when completed, GAP will reduce the flow of water into Iraq by approximately 80 per cent and into Syria by about 40 per cent. This spells bad news for Iraq, as it has historically been the substantial user of these water resources. With Syria also beginning to construct dams along the Euphrates, declining flows into Iraq have reached a point of contention.

It is highly unlikely that Turkey will discontinue its ambitious GAP project. Once finished, it is likely to withdraw up to 70 per cent of the Euphrates' water, which is likely to lead to a further deterioration in Iraq's water share which is already in a constrained situation and further exacerbate the strained relations between the two countries. Iraq threatened to take its case for an increase in water flows from Turkey and Syria to the UN. This may be a precursor to an increasingly aggressive stance by Iraq towards its upstream neighbours. Tension is high amongst the people of Iraq but as of now the Turkish government seems highly adamant and not easily willing to abandon the construction of conceived dams.

At present, out of 22 dams envisaged under the GAP project, 12 have been completed, 2 are ongoing and 8 are in the planning stage. Out of 19 Hydro Electric Energy Power Plants (HEEPPs), 6 have been completed so far and construction on 2 is going on. The rest have yet to be taken up.

Tigris and Euphrates Rivers

Both the Tigris and Euphrates originate in Turkey and pass through Syria and Iraq. These rivers have experienced drastic reductions in water flows in recent years owing to massive Turkish hydro-engineering undertakings and continual regional droughts, resulting in ominous consequences for Iraq, which has historically prospered because of the rich agricultural harvests based on water supply sourced from these two rivers. Turkish ventures aimed at massively expanding their exploitation of the water of the two rivers have coincided with severe droughts in the region and have resulted in escalating water crisis in Iraq. This problem borders on an environmental upheaval. Political negotiations between the three countries have so far fallen short of reaching any amicable solution.

Heightened Iraqi Woes

The gradual decline in water flows in the lower Tigris and Euphrates has led to the seeping of salt water from the Persian Gulf into the Shatt Al-Arab River in Iraq. Iraq is ready to accept levels upto 1,500 parts of salt per million—which is highly saline yet suitable for date palm agriculture—but recorded levels go upto 40,000 parts per million in 2009, at the peak of a severe drought, and 12,000 parts per million since then. The situation has deteriorated so significantly, that in 2009 the salt water reached beyond Basra, Iraq's biggest port city, to Qurna, nearly 150 km inland at the confluence of the Tigris and Euphrates. This decimated agriculture, that largely relies on the waters from the Shatt Al-Arab, as well as freshwater fisheries on the river. Crops, livestock and the traditional groves of date palms in this area have all had to be abandoned by farmers. This has led to the forced migration of around 30,000 people from the southern marshlands northwards. A 2009 Unesco report found that 100,000 Iraqis have fled since 2005 due to water scarcity. If this trend continues, regional tensions in Iraq are likely to further escalate, as areas already stressed by faltering economic development will be forced to deal with an influx of citizens from areas ruined by salinity and water shortages.

Raging Controversy over Ilisu Dam

Ilisu Dam is the largest project in Turkey's ambitious GAP. With designed potential of 1,200 mw, it is being constructed on the Tigris River in Southeast Turkey, some 50 km from the borders with Syria and Iraq. An 1,800 m-long wall will form a 313 sq km reservoir. The original plans date back to the 1950s and the actual design was approved in 1982.

The Ilisu dam project will displace around 78,000 mostly Kurds and simultaneously flood the 12,000-year-old city of Hasankeyf. It is an area that meets nine out of 10 Unesco criteria for World Heritage status. In November 2019, artefacts and a fifteenth century tomb were moved out for relolcation by the Turkish government.

When the dam's outlines were finalised in the 1980s, it quickly prompted strong reaction from local communities, environmental, cultural heritage and human rights groups as well as academics and celebrities in Turkey and abroad. In 2009, three European credit agencies withdrew from the project due to the Turkish government's failure to meet international standards to protect Nature, culture and the rights of tens of thousands who would be displaced.

Resettlement and Rehabilitation Issues

The southeast of Turkey is inhabited predominantly by ethnic Kurds. A civil war between the Kurdish Workers Party (PKK) and Turkish security forces killed around 35,000 people from 1984 to 1999 and led to the displacement of hundreds of thousands more. Armed clashes continue and human rights violations and restrictions on the freedom of expression are not uncommon. Tens of thousands of chiefly Kurdish origin people may be affected by the Ilisu dam project. The resettlement plan put forward by the project sponsor is utterly flawed. No deliberations with the affected population have taken place. The availability of resettlement land is still obscure, income restoration measures are vague and have not been discussed with the people. Mayors of the surrounding cities have already declared that they cannot cope with the influx of thousands of ousted. The impoverishment of a great part of dislodged people and increasing social tensions must be expected. The acquisition of the initial villages close to the dam construction site started in mid-2007. Affected people were not offered any suitable resettlement land and received only minimal cash compensation. The relocation of the village of Ilisu, which took place at the end of 2010, confirms the concerns. People are neither allowed to grow vegetables nor allowed to keep livestock, depriving them of their traditional livelihood sources.

Downstream Riparian Neighbours' Plight

It is quite evident repercussions would be felt downstream of the dam. If seen in conjunction with the smaller Cizre dam, planned between Ilisu and the borders with Syria and Iraq, inflows to the downstream riparian states may fall to drastically low levels. The Iraqi government has articulated great concern over the project. Iraqi people are also in deep anguish over the issue and have created pages on social networking sites as well to draw global attention. Apart from this, in contravention of the internationally accepted laws and framework, both Syria and Iraq have neither been fully informed nor consulted on the critical issues of the project.

Source: google map used in Geographical.co.uk

Inconsistencies with Accepted International Standards

The lack of consultation with neighbouring countries is a clear breach of international water law and it is also likely that the financial institutions supporting the project may be held accountable. Although the consortium, as well as the export credit agencies, have announced they will only proceed with the project if international standards are met, both the environmental impact assessment (EIA) and the Resettlement Action Plan (RAP) fall far short of international standards.

The export credit agencies (ECAs) supporting the project imposed some 153 conditions on its sponsors. Despite filling some of the worst gaps between Turkish law and international standards, these conditions still fail to bring the project in line with international standards. Moreover, an analysis of the 153 conditions attached to the export credit guarantees undertaken by Berne Declaration, ECA-Watch Austria, CounterCurrent, The Corner House and FERN finds that the project approval process is in stark breach of World Bank standards. The terms of reference (TOR) are vague and of poor scientific quality. The irony is that the project implementation itself neither complies with the ECA's own conditions nor with World Bank standards.

A submission by CounterCurrent and the Initiative to Keep Hasankeyf Alive to the UN Committee on Economic, Social and Cultural Rights concludes that the project implementation and the Turkish laws on expropriation and resettlement lead to severe infringements of the Covenant on Economic, Social and Cultural Rights. More cases are to be expected to crop up if the project goes ahead.

Extensive Unrest Spreading in the Area

According to various surveys undertaken so far, four-fifths of the affected populations actively resist the project. The 'Initiative to Keep Hasankeyf Alive' consists of professional chambers (architects, engineers, lawyers), elected representatives of affected communities from various political parties, human rights and environmental organisations. All have been resisting and fighting against the dam project. Around 2,000 joined a demonstration in Hasankeyf demanding the cancellation of the project on 20 May 2007. In October 2010, hundreds of participants took part in a protest camp in the ancient town of Hasankeyf. The Turkish Nature Association Doga Dernegi has been raising awareness all over the country of the environmental and cultural impacts of the Ilisu dam. Celebrities like the famous pop singer Tarkan and Nobel Prize winner Orhan Pamuk pledged themselves to the protection of the ancient town. Thousands of clients have cancelled their accounts with Garantibank, including the collective withdrawal of the Turkish labour union Petrol-Is' accounts.

INDUS SYSTEM OF RIVERS AND INDUS WATER TREATY: RESENTFUL FRETFUL PAKISTAN

Background of Indus Water Treaty

At the dawn of independence, the boundary line between the two newly created independent countries, Pakistan and India, was drawn right across the Indus Basin, leaving Pakistan as the lower riparian. Moreover, two important irrigation headworks, one at Madhopur on Ravi River and the other at Ferozepur on Sutlej River, on which the irrigation canal supplies in Punjab (Pakistan) had been completely dependent, were left in

Indian territory. A dispute thus arose regarding the utilisation of irrigation water from existing facilities. Negotiations held under the good offices of the World Bank culminated in the signing of Indus Waters Treaty in 1960. The Treaty has been effective since 1 April 1960.

The treaty's provisions could be termed as most generous of treaties in the latter half of the twentieth century. The important clauses include:

- Constitution of the new permanent post of Commissioner for Indus Waters and formation of the Permanent Indus Commission (PIC), entrusted with the implementation of the treaty meant for the Indus system comprising three Eastern Rivers (Ravi, Beas and Sutlej and their tributaries) and three Western Rivers (Indus, Jhelum and Chenab and their tributaries).
- Allocation of the waters of eastern rivers to India put India under obligation to let the waters of the western rivers flow except for domestic use, non-consumptive use, specified agricultural use and generation of hydro-electric power as specified
- India permitted to construct storage of water on western rivers upto 3.6 million acre feet (maf) for various purposes.
- India has been permitted agricultural use of 7,01,000 acres over and above the Irrigated Cropped Area (ICA) as on 1 April 1960. Out of this additional ICA of 7,01,000 acres, only 2,70,000 can be developed (i.e. a total ICA of 9,12,477 acres including that on 1 April 1960) till storages are constructed and 0.5 maf of water is released there from every year.

Pakistan raises its voice intermittently whenever India constructs any dam on Western Rivers, despite India's legal right of making impoundment upto 3.6 maf as enshrined in the Indus WatersTreaty. Let's examine the case of Baglihar Dam a run-of-the-river (ROR) power project on the Chenab river (Western River as per Indus treaty) in Jammu and Kashmir. The project was conceived in 1992 and the construction began in 1999. Soon after, Pakistan claimed that design parameters of Baglihar project violated the Indus Waters Treaty though the treaty contains provisions for India to establish run-of-river (ROR) power projects with limited reservoir capacity and control flow needed for feasible power generation from Western Rivers. Pakistan argued that some design parameters were too lax and provided India with excessive ability to accelerate, decelerate or block flow of the river, thus putting India into a strategic leverage in times of political tension or row. During 1999-2004, India and Pakistan held several rounds of talks on the design of projects but could not reach an agreement. After failure of talks on 18 January 2005, Pakistan raised several objections to the World Bank. In May 2005, World Bank appointed Prof. Raymond Lafitte, a Swiss civil engineer, to adjudicate the difference.

Lafitte declared his final verdict on 12 February 2007, in which he upheld some minor objections of Pakistan, declaring that pondage capacity be reduced by 13.5 per cent, height of dam structure be reduced by 1.5 metres and power intake tunnels be raised by 3 metres, thereby limiting some flow control capabilities of the earlier design. However, he rejected Pakistan's objections on height and gated control of spillway, declaring these conformed to engineering norms of the day. India had already offered Pakistan similar minor adjustments for it to drop its objection. The key issue that any dam constructed by India should be strictly ROR was rejected outright. Pakistan government expressed its dissatisfaction over the final outcome.

Pakistan similarly voiced its concerns over Kishanganga Hydroelectric Plant (KHEP) in the International Court of Arbitration. The Kishanganga plant has an installed capacity of 330 mw, designed to divert water from the Kishanganga river to a power plant in the Jhelum river basin. It is located 5 km north of Bandipore in Jammu and Kashmir. Construction was halted by the Hague Court of Arbitration in October 2011 due to Pakistan's protest of its visualised effects on the flow of the Kishanganga river (Neelum in Pakistan). In February 2013, the court in its interim order maintained that India could divert a minimum amount of water for power generation. Pakistan is concerned that the project would have adverse impacts on the flow of the river into their country.

This project is now operational and the Prime Minister of India inaugurated it in May 2018. Yet Pakistan continues to register its complaint in some or other international forum.

Pakistan is also constructing the Neelum–Jhelum Hydropower Plant downstream of the Kishanganga. It will divert a portion of the Neelum River from Pakistan which will reduce power generation at the Neelum–Jhelum Hydropower Plant. India argues its project will divert only 10 per cent of the river's flow. Nevertheless, water flow below the Neelum–Jhelum Dam, in Pakistan's Neelum Valley, is expected to decline as both projects are diverting water of the Jhelum. In 2010, Pakistan appealed to the Court of Arbitration, complaining that the KHEP violates the Indus River Treaty. In February 2013, the Hague ruled that India could divert a minimum of water for the project. In this partial award, the court upheld

India's main contention that it has the right to divert waters of Western Rivers, in a non-consumptive manner, for optimal generation of power. The court gave its 'final verdict' on 20 December 2013, wherein it allowed India to go ahead with the construction of the Kishanganga dam over which Pakistan had raised objections. The court delivered its 'final award' after India requested clarification of an order issued by it in February. The final award stipulates maintaining 9 cum of natural flow of water in Kishanganga river at all times to sustain the requisite flow downstream. The court also upheld alternative techniques would have to be deployed for Kishanganga hydroelectric project and all future ROR projects undertaken on Western Rivers of the Indus system.

The latest case under dispute between India and Pakistan is the construction of the 45-mw Nimoo-Bazgohydropower project on the Indus by India. This ROR power project is situated at village Alchi, 70 km from Leh. The project was conceived in 2001 and construction began in mid-2005. Pakistan first made a fuss over carbon credits won by this project and later on the federal government decided not to file a lawsuit in the Permanent Court of Arbitration-International Court of Arbitration (PCA-ICA) in Hague regarding its concerns and grievances over the Nimoo-Bazgo hydroelectric power project.

As of now, Pakistan is raring to raise its concern on the four new projects of India which include Ratle Hydroelectric Plant (48 mw), Miyar (120 mw), Lower Kalnai (48 mw) and Pakul Dul (1,000 mw to be ultimately increased to 1,500 mw). Pakistan reportedly has serious objections to the design of Ratle Hydropower Project and minor ones on the remaining three projects.

WATER DISPUTES IN MIDDLE EAST

Former UN Secretary General Boutros Boutros-Ghali once proclaimed in the 1990s that the next war in the Middle East would be about water not politics, still the possibility of future war over water is not ruled out outright.

Six-Day War

Many have described the 1967 Six-Day War as the first modern water war, resulting from clashes between Israel, Jordan and Syria over competing claims to the flow of the Jordan river. Syria attempted to divert some of the river's headwaters in collaboration with Lebanon and Jordan. The war's upshot treated the region's water as spoils of war to be distributed among contestants, with the lion's share going to the conqueror.

The Jordan river traverses from the Sea of Galilee to the Dead Sea, forming a natural boundary as well as separating Israel from Jordan. It originates from several sources, mainly the Anti-Lebanon Mountains in Syria. It then flows down into the Sea of Galilee and finally drains into the Dead Sea. This 251-km long river passes through three countries: Israel, Palestine and Jordan. Its main tributary is Yarmouk on the left bank. Other small tributaries on the left are rivers such as Banias, Dan and Zarqua. On the right side it is joined by Hisbani river (Lebanon) and Iyon.

The main bone of contention is the pumping station which Israel commissioned on the river in 1964 some 3 km from the source. Downstream from the pumping station, the river flow is now considerably reduced, rendering the river a puny creek. The reduction of water of the Jordan river is a cause of the struggle for water that has shaped the modern Middle East.

The flow of a river that once irrigated the fields of the West Bank has been siphoned off through pipes, pumps and canals to cater to taps in Tel Aviv, and to let the Negev desert bloom. This technical marvel of diversion is not only emptying rivers but also rendering aquifers dangerously susceptible to the intrusion of salt water and raw sewage.

Two well known areas—Golan Heights and the West Bank—both contain important aquifers. Israel wrested Golan Heights from Syria in 1967 and finally annexed this territory in 1980. West Bank was captured from Jordan. Syria wants Israeli withdrawal to 5 June 1967 borders, allowing Syria access to the Jordan and Yarmouk rivers. Israel wants to use boundaries dating back to 1923 and the British Mandate, which give the areas to Israel.

On the other side of the borders, both Jordan and Syria have snatched up what they could by means of building dams, digging wells and diverting as much water as possible. Over the decades, each nation has usurped the water in the areas they control following battles. The 1994 peace treaty between Israel and Jordan mandated provisions for managing shared water resources—but it unfortunately leaves little to the Palestinians who live alongside.

Prolonged skirmishes and years of sabre-rattling resulted in Israel quadrupling the territory it once controlled and gaining absolute control of double its fresh water resources. Thus it emerged a clear winner of the water wars!

Palestinians often argue water politics is just part of the injustice of Israeli occupation. They also contend they are prevented from using their own water resources by a formidable military power, forcing many to buy water at inflated rates.

DAMMING OF BRAHMAPUTRA BY CHINA

Since Brahmputra's river basin area (5.8 lakh sq km) is spread over four countries—Tibet, Bhutan, India and Bangladesh—it adds an international dimension to this river. The world has been keenly watching China's over-indulgence in this lucrative basin, especially in upper Brahmaputra basin. Out of this total catchment area of 5.8 lakh sq km, 50.5 per cent lies in Tibet, 33.6 per cent in India, 8.1 per cent in Bangladesh and 7.8 per cent in Bhutan.

This amazingly beautiful river virtually resembles a moving ocean at various junctures and acquires mind-boggling width at select locations. In Upper Assam near Dibrugarh, the river widens to 16 km whereas in lower Assam at Pandu near Guwahati, the river is just 1.2 km wide. In the immediate downstream, it stretches as wide as 18 km, unmatched on the entire planet!

The name Brahmaputra means 'Son of Brahma', one of three eternal gods of Hindu mythology. It has many other names with change of geography: Tsangpo (Tibet), Yarlung Zangbo (Chinese name), Siang/Dihang (Arunachal Pradesh) and Jamuna (Bangladesh). It has earned a few other epithets like 'Moving Ocean' and 'The River of Battleground'.

The Brahmaputra river is one of the youngest largely free-flowing rivers emanating from the Tibetan plateau which is also called Earth's third pole. Sometimes dubbed 'Water Tower of South Asia', the Brahmaputra flows for about 1,625 km inside the Tibet Autonomous Region of China and a further 918 km inside India. Thereafter, it runs 337 km in Bangladesh before finally emptying into the Bay of Bengal.

It is learnt from reliable sources that China has put

on temporary hold its overly ambitious plan of diverting Brahmaputra's massive waters to its parched north mainly because of 'Law of Gravity Flow' posing a deterrent to its audacious venture. However, it seems to be bent upon building a series of hydro-electric projects to harness electricity out of this mighty river of mammoth potential.

China has been building the Zangmu Hydropower Station on the main stream of the middle reaches of Brahmaputra, the first of a series of planned dams (Lengda, Zhongda, Langzhen, Jiexu, Jiacha and Zangmu). Zangmu Water Power Station has designed capacity of 510 mw. The project is located in Lhokha (Shannan) Prefecture of Tibet Autonomous Region, about 140 km southeast from Lhasa. On 12 November 2010, the builders of the Zangmu Hydropower Station announced the successful damming of the river. Downstream riparian countries like India and Bangladesh will likely be severely impacted.

In 2013, China's State Council proclaimed its new Energy Development Plan in January, which included several controversial dams that had previously been temporarily, abandoned owing largely to a host of environmental concerns and public opposition. Among these were three new hydropower dams on the middle reaches of the Brahmaputra river: the 640 mw Dagu Dam (which lies 18 km upstream of Zangmu), the 320 mw Jiacha Dam (also on the middle reaches of the Brahmaputura downstream of Zangmu), and the Jiexu Dam (11 km upstream of Zangmu).

Dams of Discord: Indian Fears

The three dams—Jiexu, Zangmu and Jiacha—are roughly 25 km of each other and about 550 km from India's border. The first one has been independently confirmed to be an ROR project which will not significantly impound water in a large reservoir. Construction on the second in the series, Zangmu, began in 2010. However, Indian authorities are sceptical whether this will be a pure ROR. The third, a 320-mw dam, intended at Jiacha, approximately 12 km downstream of Zangmu, is also said to be ROR.

These are not the only dams about which India has not been sufficiently informed. A dam near Zhongda and another near Phudo Zong, as well as a number of other projects were planned and executed by Beijing previously, revealing little to India.

Motuo Hydro Projects in Tibet: Another Three Gorges

Some reputed and reliable sources indicate an exercise on another ultra-mega hydro-elecric project on the Brahmaputra: Motuo on the great bend of the Yarlung Tsangpo in Tibet as part of a huge expansion of renewable power in the Himalayas. Tapping the power of the river as it bends and plunges from the Himalayan roof of the world down towards the Indian and Bangladeshi flood plains has long been a dream of Chinese hydro-engineers.

Various documents on the website of a government agency (Hydro China – a governmental enterprise for dam building) suggests a 38 gw hydropower plant is under consideration that would be more than half as big as the Three Gorges dam. However, the Chinese government as usual does not confirm the existence of such a scheme.

The mega-facility is among more than 28 dams on the river that are either planned, completed or under discussion in China. Motuo project would involve the construction of a series of tunnels, pipes, reservoirs and turbines to exploit the spectacular 2,000 metre fall of the river as it curls down towards India. A large dam such as Motuo would be tantamount to a major, irreversible experiment with geo-engineering. Blocking the Yarlung Tsangpo could devastate the fragile ecosystem of the Tibetan Plateau, and would withhold the river's sediments from the fertile floodplains of Assam in north-east India, and Bangladesh. It would also severely impact riverine ecology and induce geomorphological and ecological modification in immediate and far downstream area.

XAYABURI DAM ON RIVER MEKONG

The Mekong river is one of the world's great river systems, flowing 4,909 km through six countries: China, Myanmar, Thailand, Lao PDR, Cambodia and Vietnam. The source of the river's great productivity is its seasonal variation in water level and the range of wetland habitats inundated. The Mekong river basin's bio-diversity is immense, even in comparison with other parts of tropical Asia. Its biodiversity is fundamental to the viability of natural resource-based rural livelihoods of a population of 60 million people living in the lower Mekong basin.

This river has mammoth hydro-electric power potential, roughly around 60,000 mw. At least 30 dams have already been built on various locations spread over all across the Mekong basin countries and over 130 are in the pipeline.

The river is also endowed with immense diverse natural resources. Among the world's river basins, only the Amazon possesses a greater diversity of plants and animals. On an average, an astounding 15,000 cum/second) of water flows into the Mekong mainstream from the surrounding basin area—enough to meet the daily needs of 100,000 people. The water irrigates large tracts of forest and wetlands that produce building materials, medicines and food and serve as habitats for thousands of species. The river Mekong provides food security, livelihood and cultural identity to the people living in China, Cambodia, Laos, Thailand and Vietnam.

Mekong River Commission

Way back in 1959, the UN constituted the Mekong Committee with headquarters at Bangkok, Thailand. Its successor, the Mekong River Commission (MRC) came into existence in 1995 leading to the Mekong Agreement. The responsibility of the management of the MRC is in the hands of its four member countries: Cambodia, Laos PDR, Thailand and Vietnam. It holds regular meetings and summits. The first MRC summit was convened in Thailand in 2010 and the second in April 2014 at Ho Chi Minh City, Vietnam

The Laos government with the active help of the Thai government commenced building the massive 1,285 mw Xayaburi Dam on the Mekong in 2010. The project is being constructed by Thailand's prestigious construction company Karnchang and financed by six Thai commercial banks, including the state-owned Krung Thai Bank. EGAT, Thailand's electricity utility company, has agreed to purchase 95 per cent of the installed capacity electricity.

The costs of the Xayaburi Dam will be hugely borne by the hundreds of thousands of riparian people along the Mekong River, including in Laos and Thailand. It is largely held that the dam will block critical fish migration routes for nearly 100 species, including the Mekong Giant Catfish. The dam would also decimate the river's complex ecosystems that serve as important fish habitats. It would block the flow of sediments and nutrients, affecting agriculture as far downstream as the Mekong Delta in Vietnam. At least 2,100 people would be displaced and over 200,000 people living near the dam site will be directly impacted.

As of now, construction on the project has been proceeding rapidly and much of the river has already been blocked off. However, at the MRC council meeting in January 2013, Cambodia, Vietnam and the MRC's donor governments all articulated their concern over the said project.

There was a breach in this dam in August 2018 but work is continuing despite resistance.

Contentious Issues

The construction is ruffling feathers in all the four Mekong river basin countries of Cambodia, Vietnam, Laos and Thailand. The main conflicting issues are:

- According to a 1995 treaty, the four governments of Cambodia, Laos, Thailand and Vietnam are required to seek agreement before building any dam on Mekong river. The process is called 'prior consultation' and is to be facilitated by the MRC. However, such an agreement was never sought by Laos government for the construction of Xayaburi Dam
- Both Laos and Thailand are refusing to study the dam's trans-boundary impact despite requests by Vietnam and Cambodia
- The project's current environmental impact assessment only examines impacts upto10 km downstream of the dam site
- The full extent of the dam's impact remains unknown.

BRAZIL'S BELO MONTE DAM

Despite widespread unrest fermenting in the Amazonian basin, the Brazil government has been moving ahead with this highly controversial dam. Upon completion, it would be the world's third largest hydro-electric dam on one of the Amazon's major tributaries, the Xingu river. The Belo Monte Dam complex is designed to divert 80 per cent of the Xingu river's flow, severely affecting an area of over 1,500 sq km of Brazilian rainforest while resulting in the forced displacement of 20,000-40,000 people. The project is causing grave threats to the land and livelihood of thousands of riverine and urban families, as well as 1,000 indigenous people.

The project is highly complex in nature and intertwines two dams–Pimental Dam (233 mw) and Belo Monte Dam (11,000 mw). Both envisage house turbines and spillways. The Pimental Dam will block the Xingu, diverting up to 80 per cent of the river's flow to the main reservoir, while the spillway will release a minimum of 20 per cent of its flow to the 60-mile Big Bend or Volta Grande. The Belo Monte Dam will receive the water from the main reservoir, while its spillway will release this water back into the lower Xingu river. This way two reservoirs shall be created: one in the Xingu riverbed, and the other on dry land. A total of 668 sq km would be flooded, including 400 sq km of forest; in all, at least 1,522 sq km will be directly affected. Apart from this, one massive canal 500 metres wide, and a series of dykes to transfer up to 80 per cent of the Xingu's flow into an artificial canal have been planned.

The project has been encountering fierce resistance from the Xingu's indigenous peoples and social movements for over 20 years. Though all the turbines of the project were online by

2019, the social and environmental impacts remain intact. The desired water for producing electricity and meeting turbine head is not coming in the reservoir. Resentment continues, with people calling it not Belo Monte (Beautiful Hill) but Belo Monstro (Beautiful Monster). It was intended to attain 11,233 mw operational capacity per month but has not exceeded 6,882 mw. It poses very serious threats to human life and globally important ecosystem as per expert testimony.

Independent studies have confirmed the dam's social, environmental, technical and economic non-viability. Simultaneously, the project's legality has been challenged through a series of lawsuits that have reached Brazil's Supreme Court. Despite all this, the project has advanced significantly in the last two years, receiving government approval to initiate construction and obtain unprecedented public financing from Brazil's National Development Bank. Indigenous communities are worried about their future and continue to vent their resentment and anguish through mass demonstrations, marches and week-long protests. Many international organisations like International Rivers and Amazon Watch have been espousing the cause of affected people and aggressively building international opinion against the ongoing dam.

FUTURE COURSE OF WATER CONFLICTS

The world is in the throes of dire crisis in several regions. The Middle East, North Africa and South Asia are all destined to experience severe water shortages over the coming years mainly because of decades of mismanagement, outdated policies, insensible wastage and ignorant overuse.

The global scramble for water will lead to new conflicts

and basin disputes. We are standing on a cliff looking over the edge and we have to decide what we are going to do.

One US intelligence report warns as water shortages become more acute in the next decade, water in shared basins will increasingly be used as leverage; the use of water as a weapon or to further terrorist objectives will become more likely beyond ten years.

It remains to be seen how intelligently we address the precipitating crisis, resolve people's fears, manage wisely our existing water resources, effect fair and rational water pricing and sensitise people for the prudent use of this precious commodity.

The Global Hankering for Gas and Gasoline

THE YEAR 1982 was a watershed year when global oil consumption exceeded oil production. Global fundamentals are rapidly driving even more robust demand for both oil and gas. Between 2010 and 2040, global energy demand was projected to increase by one third reaching 120 bbl of oil equivalent per annum. This, it was thought, would be driven by a projected rise in global population from 7 billion to 9 billion and growth in the global economy of 130 per cent. Brent oil prices have risen from $20 per barrel at the turn of the millennium to $110 per barrel in recent years. However, due to the impact of Covid-19 worldwide, IEA has said that oil demand could drop by 9% by the end of 2020 but the demand for renewables may go up because of low operating costs and preferential access to many power systems. Brent price dramatically fell to $18.38 per barrel in the month of April 2020 due to world-wide lockdown.

Until recently, the vast majority of the world's oil came from conventional sources, wells drilled on land or not far offshore. However, as conventional production declines over the next few decades, more of the world's oil demand will be met by emerging sources such as oil sands, tight oil, deepwater (wells drilled in more than 400 metres of water), natural gas liquids and to some extent bio-fuels. Global energy market today represents a strong sector with dynamic growth inviting international players in the action packed arena of oil and gas!

It becomes essential to trace gas (natural gas) and gasoline (petrol) and other hydrocarbon resources the world is so desperate about. Millions of years ago, the remains of plants, vegetation, diatoms (single-cell algae with a wall of silica) and animals decayed, built up and amassed in thick layers. This decayed matter from plants and animals is called organic material. As time passed by, the sand and silt changed to rock, covered the organic material and trapped it beneath the rock. Action of both pressure and heat changed some of this organic material into coal, some into oil (petroleum), and some into natural gas. The main ingredient in natural gas is methane, a gas or compound mainly composed of one carbon atom and four hydrogen atoms. Oil and gas are held in 'reservoirs' which connote geological formations of porous and permeable rocks that can hold significant amounts of oil and gas within their pores.

Proved Reserves and Undiscovered Technically Recoverable Resources

Proven reserves of natural gas are estimated quantities that analyses of geological and engineering data have demonstrated to be economically recoverable in future years from known reservoirs.

Proved reserves are added each year with successful exploratory wells and as more is learnt about fields where current wells are producing. For this reason, those reserves constantly change and should not be considered finite resources. Application of new technologies can convert categories of previously uneconomic natural gas resources into proven reserves.

In addition, there are large volumes of natural gas classified as undiscovered technically recoverable resources. These are expected to exist because geologic settings are favourable, despite the relative uncertainty of their specific location. They are also assumed to be producible over some time period using existing recovery technology.

Growing Global Natural Gas Consumption

Commercial use of natural gas has gained immense popularity in the world as it is one of the cleanest, most convenient and economical of fuels. In some countries, it is liquefied (called LNG) and transported through cylinders to the consumers whereas in some geographies it is piped underground directly to each property, obviating the need to order or store it, always available when needed. Over the past 10 years, natural gas has proven itself to be the most cost-effective fuel in the energy space. It finds application in a number of ways including cooking, central heating, natural gas fires, tumble drying, barbeques, space heating, combined heat and power (CHP), *et al.*

In the past one decade, the global consumption of natural gas has risen significantly which was just 99.6 trillion cubic feet (tcft) in 2005 and climbed to 117 tcft in 2013. Global consumption is likely to rise further in coming decades. In 2018, global consumption figures touched 139.67 tcft levels and are likely to zoom further to 185-191.4 tcft in 2040. Figures of 191.4 tcft have been envisaged in the high oil price scenario which is more likely to persist owing to a number of contributing factors viz. fluid and fragile Middle East geopolitical situation, considerate and conservative oil withdrawal by the OPEC countries, changing global political equations in central Asia with Crimea's annexation by Russia, slow pace of growth of alternative fuels, global resistance and scepticism surrounding shale gas extraction.

The linear charts depicted below (Graphics 6.1 and 6.2) explicitly show the yearly growth pattern in global natural gas consumption. It is evident the world is gradually trending towards higher consumption. Data of the past fourteen years (2005-2018) retrieved from USEIA and Global Energy Statistical Yearbook 2019 clearly corroborate the consistent rise in global consumption. It rose from 100 tcft in 2005 to around 140 tcft, an increase of 40 per cent.

Main drivers of global gas consumption are the US and China, which accounted for around two-thirds of the additional consumption. US gas demand grew by 10 per cent in 2018, the

Graphic 6.1: Global Natural Gas Consumption 2005-2013

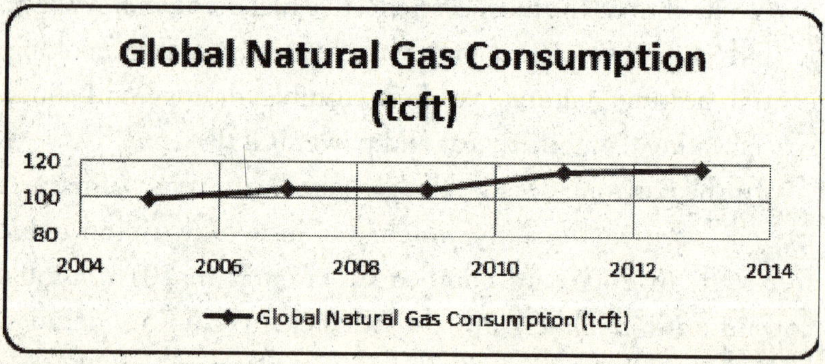

*Data sourced from USEIA

Graphic 6.2: Global Natural Gas Consumption 2014-2018

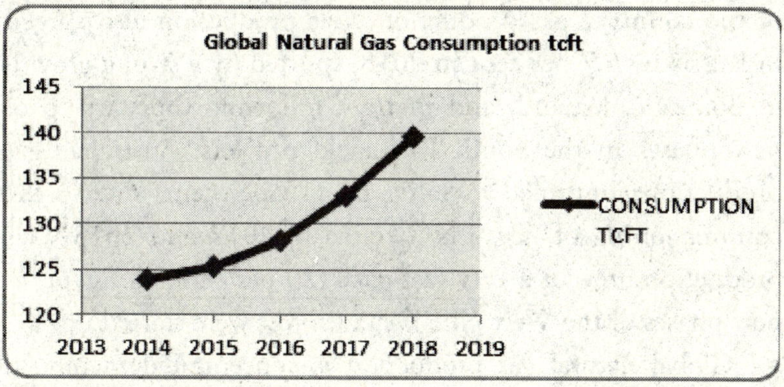

*Data source Global Energy Statistical Yearbook 2019

highest growth seen in the past three decades, largely spurred by the power sector (+15 gw of new gas-fired power plants) and by buildings. Gas consumption also accelerated in China (+18 per cent) in line with its recent coal-to-gas substitution policy in the power and heating sector. Escalation was also seen in India and South Korea owing to sustained economic growth. Gas consumption also continued to grow steadily in Russia though at a slower pace than in 2017 and accelerated in Canada, Iran and Algeria.

Global Natural Gas Production Import and Export

Global extraction of natural gas was around 2,000 bcum in 1990 which has now catapulted to nearly 4,000 bcum (139.67 tcft) in 2018, just about doubling in a span of 28 years!

The global surge in gas production in 2018 has largely come from the US, accounting for 45 per cent of the global increase. Gas production soared in the United States by 11.5 per cent, pushed by development of the Permian Basin and Haynesville shale formations.

Shale gas in the US now accounts for around 70 per cent of the country's gas production. Gas production also surged in Russia by 6.7 per cent in 2018, spurred by a strong growth in domestic demand, and in Iran, following the start-up of new phases in the South Pars Field projects. Australia's gas production continued to ramp up (15 per cent) due to the commissioning of new LNG trains in 2017 and 2018. Gas production grew at a very fast pace (20 per cent) in Egypt, as new phases of the West Nile Delta project were started up.

Global natural gas production is concentrated mainly in five countries (see Graphic 6.3) namely USA, Russia, Qatar, Iran and Canada. These five countries alone contribute more than 50 per cent of the world production. Moreover, USA and Russia together command nearly two-fifths of the entire global production. Other countries that share 2-4 per cent share are Norway (3.3 per cent), China (3.1), Saudi Arabia (2.8), Netherlands (2.3) and Indonesia (2.2).

It is also worthwhile to look at the Top 10 natural gas importers and exporters. As per the data released by the Paris-based International Energy Agency (IEA), the top exporters in 2012 were Russia, Qatar, Norway, Canada, Algeria, Turkmenistan, Indonesia, Netherlands, Nigeria and Malaysia in descending order. Top importers in the same year comprised of Japan, Germany, Italy, Korea, Turkey, USA, France, UK, China and Ukraine.

Crude Oil, gasoline and Other Petroleum Products

A layman is often confused by the terms crude oil, petrol, petroleum, hydrocarbons and gasoline. It is, therefore imperative to give a brief on the distinctions among them.

Graphic 6.3: Global Natural Gas Production Period 2014-2018

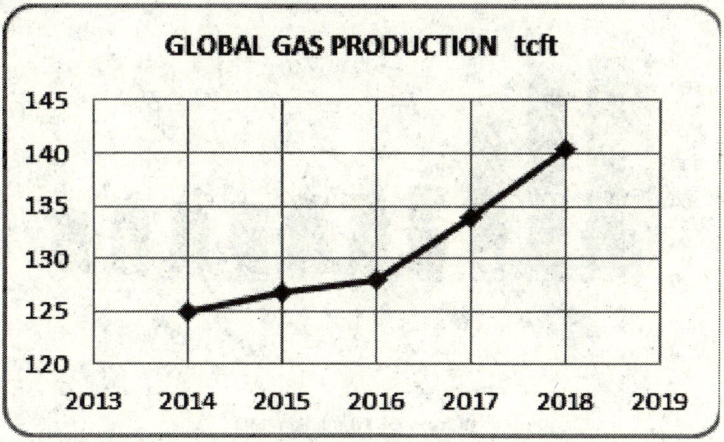

*Data source Global Energy Statistical Yearbook 2019

Graphic 6.3: Global Natural Gas Production Share Year 2012

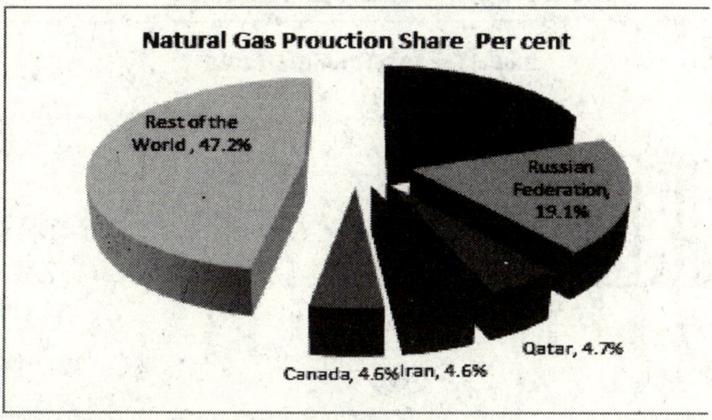

*Data sourced from IAE (International Energy Agency) Paris

Crude oil is a mixture of hydrocarbons that exists as a liquid in natural underground reservoirs and remains liquid when brought to the surface. Petroleum products are produced by processing it at refineries and the extraction of liquid hydrocarbons at natural gas processing plants. Petroleum is the broad category that includes both crude oil and petroleum

Graphic 6.3: Top 10 Crude Oil Producer Countries 2011

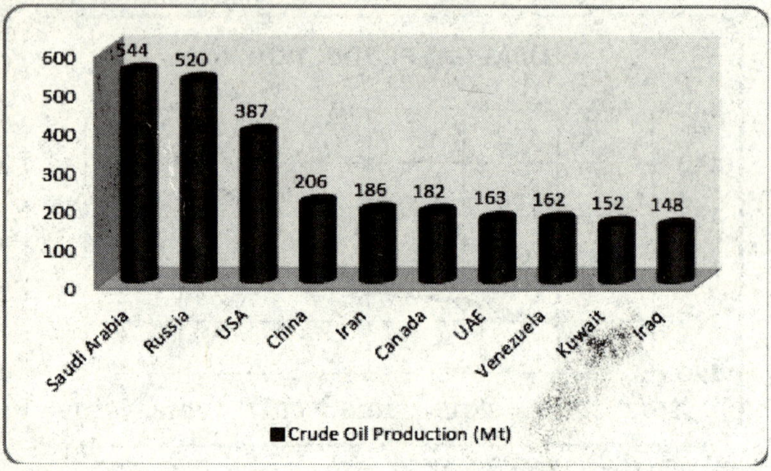

*Data sourced from IEA, Paris

Graphic 6.4: Top 10 Crude Oil Producer Countries 2018

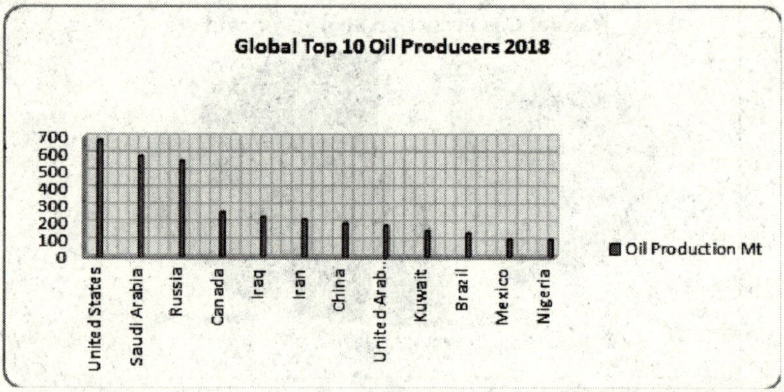

Source: US EIA & IEA

products. The terms 'oil' and 'petroleum' are often used interchangeably. The fractional distillation of crude oil renders a number of products including diesel, fuel oil (for power plants and ships), gasoline (petrol, in US also called simply gas), kerosene, naphtha and paraffin (jet fuel). Hence crude oil is the mother of all petroleum products.

One 42-gallon barrel of crude oil renders 19.4 gallons of gasoline. The rest is used to make at least 6,000 other products which includes everyday toothbrushes and toothpaste, shampoo and combs, cold creams and aspirin, even ice cube trays.

An average American consumes petroleum products at a rate of three-and-a-half gallons (13.25 litres) of oil and more than 250 cubic feet of natural gas per day. The figures are simply astounding because petroleum is not just used for fuel alone!

Now, let's have a look at the Top 10 crude oil producers in the world. Over a quarter of the total world production comes from the two countries Saudi Arabia and Russia, with 13.1 per cent and 12.6 per cent share respectively. Global production of crude oil stood at 4,142 mt in 2011. It is interesting to note that around 64 per cent of the total production is concentrated in ten countries: Saudi Arabia, Russia, US, China, Iran, Canada, UAE, Venezuela, Kuwait and Iraq. Remaining 36 per cent is contributed by the rest of the world. Here is the graphic representation of countries and their total oil produce in 2011 and 2018.

A cursory look at the two graphs reveals that the US has now occupied the top slot among the global oil producers after sticking at third place in 2011, largely driven by explosive growth in the United States by 16.5 per cent in 2018. Nigeria, Brazil and Mexico are new entrants in the club of Top 10!

In June 2018, OPEC members agreed to increase oil production to prevent supply shortage and reduce prices after previous production cuts were deemed excessive and raised prices too much. This stabilised global prices, before they rose again on anticipation of US sanctions on Iranian oil exports. Crude oil production in the US saw a significant increase as exploration and extraction of oil from its plentiful shale reserves grew, causing the largest ever annual increase by a single country. This was due to new projects coming online, a persistent demand for oil and higher prices ($14 per barrel higher than 2017).

Oil production also continued to increase in Russia, the Middle East (except Iran) and Africa. Countries like China and India produce crude oil in good quantity yet their consumption far exceeds production, with India's consumption nearly four times its production. Such countries become net importers as well as desperate seekers of alternatives and fresh resources elsewhere. India imports around 172 mt, a figure which keeps rising. For the current fiscal (2019-20) the import is likely to touch 233 mt.

Ongoing Global Quest for Oil and Natural Gas

In recent times, China's global quest for resources in general, and oil and natural gas in particular, has received unprecedented worldwide attention and global scrutiny. By the beginning of this century, it had embarked heavily upon overseas oil and gas asset acquisitions. Its overseas shopping spree spread over at least three dozen countries in Africa, Middle East, Latin America and Central Asia. China is not a significant producer of natural gas and occupies the ninth highest position among gas importers in the world. Recent reserve estimates suggest that China's natural gas resources are quite small, less than 2 per cent of recoverable reserves globally. Since 2000, production has tripled but still amounts to only 3 per cent or so of total global output. According to *Petroleum Economist*, quoting the influential Sublime China Information (SCI99), China's demand for natural gas is expected to be 10bn cum less in 2020 than it would have been in the absence of the corona virus Covid-19 outbreak (expected to have been between 300-340 bn cum).

If we cast a glance over Chinese energy consumption patterns, it becomes evident that oil is the second-largest source of energy accounting for 18 per cent of the country's total energy consumption while natural gas accounts for 4 per cent. Together they constitute 22 per cent of total energy consumption.

As of January 2014, China held 24.4 bbl of proven oil reserves; up over 0.7 bbl from the 2013 level and the highest in the Asia-Pacific region. China's total oil and liquid production, the fourth largest in the world, has risen by about 54 per cent over the past two decades and serves only its domestic market. However, the production growth has not kept pace with demand growth during this period. In 2013, China produced an estimated 4.5 bbl/d of total oil liquids, of which 93 per cent was crude oil. Most of the growth over the long term is from non-petroleum liquids such as gas-to-liquids, coal-to-liquids, kerogen, and biofuels.

China's oil demand growth hinges on several factors such as domestic economic growth and trade, power generation, transportation sector shifts, and refining capabilities. It has become imperative for China to constantly look for new energy resources, which has become one of its overriding priorities.

National Oil Companies (NOCs) of China emerged during the 1980s as the country moved productive assets into state-owned enterprises to enhance efficiency and promote competition. CNPC (China National Petroleum Corp), Sinopec (China Petroleum and Chemical Corporation), and CNOOC (China National Offshore Oil Corporation), are three major companies in the field of oil and gas, largely state owned. These companies established subsidiary firms like PetroChina (by CNPC) and listed them on the Hong Kong Stock Exchange. Because of spiralling domestic demand and peaking domestic oil production capacity, the Chinese NOCs fully adhered to the 'Go global' policy, in which the government encourages Chinese companies to invest and compete in international markets. Chinese NOCs seek to secure long-term access to energy resources by investing in resource-rich countries across the globe. China Bank for Development (CBD) has also extended hefty loans to oil and gas producing countries outside China since 2009.

CNPC commenced overseas investment in Sudan, Peru and Kazakhstan in the early 1990s. From January 2009 to December 2010, CNPC, CNOOC and Sinopec in association with other Chinese players have spent around $47 billion to acquire substantial overseas oil and gas assets.

The years 2009 and 2010 were full of intense international activities for Chinese NOCs. In April 2009, CNPC and KMG purchased 100 per cent stake in MMG, Kazakhstan for $3.3 billion. This deal gave China a strong foothold in Central Asia. Thereafter Sinopec went on to acquire Addax for $8.8 billion in June 2009. In the same month, CNPC and PetroChina purchased 96 per cent stake in SPC of Singapore for $2 billion. Two months later, Sinochem purchased 100 per cent stake in Emerald for Syrian and Colombian assets for $0.878 billion in Aug 2009. CNPC and PetroChina purchased 60 per cent of Athabasca Oil Sands Corps' Mackay River and Dover oil sands projects in Alberta, Canada in a deal worth $1.9 billion in September 2009.

Another Chinese company, Xinjiang Guanghuj Investment, purchased Kazakhstan TBM's 49 per cent share to jointly develop Zaysan block in Eastern Kazakhstan for $0.3 billion in September 2009. CIC of China purchased 45 per cent stake in Nobel Oil Group to fund Russian expansion plans and also purchased 11 per cent stake in KMG by purchasing global depository receipts in October and September 2009 respectively. CNOOC purchased partial share of Norwegian's Statoil's US assets in deepwater areas of Gulf of Mexico in October 2009. This ended the eventful year of 2009.

Greatly enthused by these successes, Chinese companies now turned to Latin America. CNOOC bought 50 per cent stake in the Argentinean oil company, Bridas Corps, which has oil and gas exploitation operations in Argentina, Bolivia and Chile for $3.1 billion in March 2010. Sinochem purchased Statoil's 40 per cent stake in Brazil's Peregrino Oilfield for around $3 billion in May 2010.

In March 2010, CNPC, PetroChina and Shell together bid for 100 per cent share of Arrow Energy (an Australian coal bed methane producer). Sinopec purchased 9.03 per cent in Canadian oil sands company Syncrude from ConocoPhillips in April 2010; this was another big deal costing $4.675 billion.

CNPC acquired 35 per cent stake of Shell's wholly owned subsidiary SPD (Syria Petroleum Development BV). SPD owns three production licences in Syria covering 40 oil fields with 23 kb/d. This deal cost CNPC $1.2-1.5 billion in May 2010. Sinopec purchased 40 per cent stake of Brazilian subsidiary of Spanish oil company Respol for a whopping $7.1 billion in October 2010.

CNOOC bought two-thirds of Tullow Oil's stake in three blocks in Uganda from French major Total SA for over a million dollars in October 2010. One month later, CNOOC acquired

33.3 per cent interest in Chesapeake's 600,000 net acres in the Eagle Ford Shale costing $2.16 billion. CNOOC in the same month acquired 60 per cent of Pan America Energy from BP (British Petroleum) costing another $2.47 billion. The year 2010 ended with successful acquisition of 18 per cent share of Chevron's Gendalo-Gehem deep water gas project in Indonesia with an investment of $0.68 billion.

India is the fourth largest consumer of oil and petroleum products after China, the United States and Japan. At the same time, it is also the fourth largest importer of oil and petroleum products. Its total oil production is nearly 980 tbl/d (thousand barrels per day) and consumption is around 3600 tbl/d whereas it annually produces 1,459 bcum of natural gas but consumes nearly 2,060 bcum, necessitating import of both products. This perennially places immense burden on the exchequer as well.

India depends heavily on imported crude oil, mostly from the Middle East. India had 5.5 bbl of proven oil reserves at the end of 2012, mostly in the western part of the country. It stood at 594.49 mt by end-March 2018. Domestic production has stagnated in recent years, and Indian national oil companies increasingly purchase equity stakes in overseas oil fields.

India's government promotes the country's refining sector. India has several world-class refineries in Jamnagar, and the refining industry is largely privately owned by RIL, Essar, etc. Natural gas serves as a substitute for coal for electricity generation in India. The country began importing LNG from Qatar in 2004 and increasingly relies on imports to meet domestic natural gas needs. India possesses 43.8 tcf of natural gas reserves as well, mostly located offshore. The two biggest state-owned oil companies, ONGC and Oil India, dominate the country's upstream gas sector. India is also keen on coal bed methane (CBM) and shale gas. Rising demand made India the world's sixth largest LNG importer in 2011.

The high degree of dependence on imported crude oil has led Indian energy companies to attempt to diversify their supply sources. To this end, Indian NOCs have purchased equity stakes in overseas oil and gas fields in South America, Africa and the Caspian Sea region to acquire reserves and production capability. However, the majority of imports continue to come from the Middle East, where Indian companies have little direct access to investment.

India through its PSUs namely ONGC Videsh Ltd (OVL), Oil India Ltd (OIL), Indian Oil Corporation Ltd (IOC), GAIL (India) Ltd (GAIL), Bharat Petro Resources Ltd (BPRL) and Hindustan Petroleum Corporation Ltd (HPCL) made significant investments for acquisition of exploration and production assets overseas in Sudan, Colombia, Venezuela, Russia, Vietnam, Syria and Brazil.

The Indian companies mentioned above have acquired the following overseas oil and gas projects over the years. Company-wise stake in each project have been depicted in the table below:

ONGC Videsh had already acquired stake in at least 33 oil and gas projects spread over 16 countries by March 2014. As of now in 2020, it has stake in 39 oil and gas projects in 19 countries including Azerbaijan (2 projects), Bangladesh (2), Brazil (2), Colombia (7), Iran (1), Iraq (1), Israel (1), Kazakhstan (1), Libya (1), Mozambique (1), Myanmar (6), Namibia (1), New Zealand (1), Russia (3), South Sudan (2), Syria (2), UAE (1), Venezuela (2), and Vietnam (2).

Country	Name of the Overseas Project/Block	Company/Participating Interest
Vietnam	Block 06.1	OVL – 45 per cent
	Block 128,	OVL – 100 per cent
Russia	Sakhalin-1	OVL – 20 per cent
	Imperial Energy	OVL- 100 per cent
Sudan	GNOP Block 1,2 and 4	OVL – 25 per cent
	Block 5 A	OVL – 24.125 per cent
	Khartoum – Port Sudan Pipeline	OVL – 90 per cent
Myanmar	Block A-1	OVL – 17 per cent
	Block A-3	OVL – 17 per cent
	Shwe Offshore Mid-Stream Project	OVL – 17 per cent
	Onshore Gas Transportation Pipeline	OVL – 8.347 per cent
Iraq	Block 8	OVL – 100 per cent
Iran	Farsi Offshore Block	OVL – 40 per cent
Libya	Block 43	OVL – 100 per cent
Syria	Block 24	OVL – 60 per cent
	Al Furat (4 PSA)	OVL – 16.67 per cent to 18.75 per cent
Cuba	Block N-25, 26,27,28, 29 and N-36	OVL- 30 per cent
	Block N-34 and N-35	OVL – 100 per cent
Brazil	Block BC-10	OVL – 15 per cent
	Block ES-42	OVL – 100 per cent
	Block BM-S- 73	OVL – 43.5 per cent
	Block BM-SEAL-4	OVL – 25 per cent
	Block BM-BAR-1	OVL – 25 per cent
	Block S-74	OVL – 43.5 per cent
Colombia	Mansarovar Energy Colombia Limited (MECL)	OVL – 50 per cent
	Block RC-8	OVL- 40 per cent
	Block RC-9	OVL – 50 per cent
	Block RC-10	OVL-50 per cent
	Block CPO-5	OVL – 70 per cent
	Block SSJN-7	OVL – 50 per cent
Nigeria	OPL-279	OVL – 23.21 per cent
	OPL 285	OVL – 32.81 per cent

Venezuela	San Cristobal Project	OVL – 40 per cent
	Carabobo – 1 Project	OVL – 11 per cent
Kazakhstan	Satpayev Project	OVL – 25 per cent
Iran	Farsi Block	OIL-20 per cent , IOC-40 per cent
Libya	Area 86	OIL-50 per cent, IOC -50 per cent
	Area 102/4	OIL-50 per cent. IOC-50 per cent
	Area 95/96	OIL-25 per cent, IOC-25 per cent
Gabon	Shakthi	OIL-45 per cent, IOC-45 per cent
Nigeria	OPL 205/OML 142	OIL-17.5 per cent, IOC-17.5 per cent
Yemen	Block 82	OIL-12.75 per cent , IOC-15 per cent
	Block 83	OIL-12.75 per cent , IOC-15 per cent
Sudan	Pipeline Project	OIL-10 per cent,
Timor Leste	Block K	OIL-12.5 per cent, IOC-12.5 per cent
Egypt	Block 3	OIL-25 per cent
	Block 4	OIL-25 per cent
Venezuela	Project Carabobo -1	OIL-3.5 per cent, IOC-3.5 per cent
Myanmar	Block-A-1	GAIL-8.5 per cent
	Block A-3	GAIL-8.5 per cent
	Development of Blocks A1 and A3	GAIL-8.5 per cent
Oman	Block-56	GAIL-25 per cent
Brazil	Espirito Santo (BM-ES-24:589,661,663)	BPRL-15 per cent
	Campos (BM-C-30: 101)	BPRL-12.5 per cent
	Sergipie (BM-SEAL-16: 349,426,497,569	BPRL-20 per cent
	Potiguar (BM-POT-16: 663,760)	BPRL-10 per cent

Oman	Block 56	BPRL-12.5 per cent
Australia	AC/P 32	BPRL-20 per cent
	WA-338-P	BPRL-8.4 per cent
	EP- 413	BPRL-27.8 per cent
	TP-15	BPRL-50 per cent
East Timor	JPDA 06-103	BPRL-20 per cent
UK	North Sea 48/2C	BPRL-25 per cent
Mozambique	Area-I	BPRL-10 per cent
Indonesia	Nunukan (Badik-I)	BPRL-12.5 per cent
Australia	WA 388 P	HPCL-8.4 per cent
Oman	Block-56	HPCL-12.5 per cent
Egypt	South Sinai	HPCL-25 per cent
	South Quseir	HPCL-25 per cent

Source: ONCG India

OVL invested more than $4 billion in acquisition of Hess Corporation's 2.7 per cent participating interest in the Azeri, Chirag and the Deep Water Portion of Guneshli Fields in the Azerbaijan sector of the Caspian Sea (ACG) and 2.36 per cent interest in the Baku-Tbilisi-Ceyhan (BTC) pipeline which was completed in March 2013. ACG is one of the world's biggest oil producing field in Caspian Sea; acquisition of an additional 12 per cent participating interest (PI) in Block BC-10, a deep-water offshore block in Campos Basin, Brazil in December, 2013; acquisition of 6 per cent PI in the Rovuma Area 1 Offshore Block in Mozambique from Videocon in January, 2014; (d) acquisition of a direct 10 per cent PI in the same Rovuma Area 1 from Anadarko in February, 2014. Apart from these, ONGC Videsh acquired two shallow water blocks in Bangladesh namely SS-04 and SS-09 in 50:50 consortium with Oil India Limited. ONGC Videsh has also been awarded two onshore blocks namely B2 (Zebyutaung-Nandaw) and EP-3 (Thegon-Shwegu) in the Myanmar Onshore Bidding Round 2013.

Another Indian company Videocon through its various subsidiaries and affiliates acquired various international oil and gas acreages in Mozambique, Brazil, Indonesia, and East Timor. Videocon Group is one of India's largest enterprises having interests in various sectors, within and out of India, including oil and gas upstream. In 2006, Videocon diversified into the international portfolio of exploration interests by successfully bidding for offshore blocks JPDA Block 06-103 in Timor Sea, Oman Block 56 in Sultanate of Oman. In 2007 along with BPCL under 50:50 joint venture, it acquired ten exploration blocks comprising of four concessions in Brazil offshore basins. Further, in 2009 it acquired a 12.5 per cent PI from Anadarko in Nunukan Block acreage in Indonesia.

Similarly another diversified Indian company Essar went on to acquire various oil and gas assets in Nigeria, Vietnam, Madagascar and Indonesia. It currently holds one exploratory block in Nigeria shallow offshore, one offshore oil and gas block in Vietnam, two onshore exploratory blocks in Madagascar, one on-land South East Tungkal block in Indonesia.

Indian giant Reliance Industries Ltd (RIL) through its subsidiary, Reliance Eagleford Upstream LP, entered into a joint venture with Pioneer Natural Resources Company to acquire a 45 per cent interest in Pioneer's core Eagle Ford shale acreage position in the US in two separate transactions.

RIL, through its subsidiary Reliance Marcellus LLC, entered into a joint venture with Atlas Energy, Inc. (now owned by Chevron Corporation) under which Reliance acquired a 40 per cent interest in Atlas' core Marcellus shale acreage position in the US. Again, through its subsidiary, Reliance Marcellus II LLC, it entered into a joint venture with Carrizo Oil and Gas, Inc. Under the transaction, Reliance acquired a 60 per cent interest in Marcellus shale acreage in central and northeast Pennsylvania, USA.

Apart from these, RIL has acquired thirteen blocks in various geographies in its international conventional portfolio, including two in Peru, three in Yemen (one producing and two exploratory), two each in Oman, Kurdistan and Colombia, one each in East Timor and Australia, amounting to a total acreage of over 99,145 sq km. RIL has also entered into a strategic partnership with BP.

No country wishes to lag behind in the ongoing rush for securing natural resources. Malaysia through its company Petronas secured exploration and production projects in at least 22 countries in South-east Asia, the Middle East, Central Asia, Latin America and Africa over the past two decades. Petronas secured its first overseas venture as an upstream operator in Vietnam in 1991. The success of this first venture spurred it to spread its wings further abroad. Today Petronas' international reserves in Africa, Southeast Asia, the Middle East and Central Asia stand at 6.56 billion boe, comprising nearly a quarter of Petronas' total reserves.

Transnational Pipeline Projects: Quest for Oil and Gas

In a bid to tap neighbouring resources, China has secured agreements with all potential nearby countries to import oil and gas from all directions. Line C of the Central Asia-China gas pipeline which was primarily designed to deliver natural gas from Turkmenistan has an annual carrying capacity of 25 bcum. Some 10 bcum of this volume will be delivered from Turkmenistan, 10 bcum from Uzbekistan and 5 bcum from Kazakhstan. Line C starts on the Turkmen-Uzbek border, runs through Kazakhstan and ends in the Xinjiang Uyghur Autonomous Region of China and further connects to China's West-East gas pipeline.

Other two lines, Line A and Line B were commissioned back in late 2009 and in October 2010, respectively. These connect to the second line of China's West-East gas pipeline and around 30 bcum is transported from Turkmenistan to China via these lines per year. CNPC of China began to purchase gas from Central Asia in 2009 after commissioning the first two lines of the gas pipeline. China has also signed an agreement with Turkmenistan to augment the supply volume to 65 bcum by 2020. The fourth, Line D, that will run through Uzbekistan, Tajikistan and Kyrgyzstan, is being constructed for this purpose.

ESPO (East Siberia-Pacific Ocean) Oil Pipeline

By the agreement of February 2009, China got access to the eastern Siberian oil of Russia through ESPO oil pipeline. In February 2009, China and Russia signed a long-term oil supply deal worth $25 billion. Through this agreement, China secured a 300 kb/d supply of Russian oil for twenty years.

ESPO runs from Skovorodino (Amur Oblast) to the Kozmino Sea Terminal covering the length of 2,046 km with annual capacity of 35 mt. ESPO transported 30 mt of oil, including 15 mt to China. It is projected that 36 mt of oil will be transported via this pipeline in future. The other key recipients of Russian oil via ESPO are Japan (31 per cent), the US (22), South Korea, Singapore, the Philippines, Thailand and Malaysia. China receives around 24 per cent of the total.

Two Pipelines from Myanmar

Chinese experts harbour the thought that hostile forces (possibly the US) could one day block the crucial passageway through Strait of Malacca and starve their country of energy. Till last year, more than four-fifths of China's oil and gas imports were transported by ship through a narrow waterway separating Indonesia and Malaysia, known as the Strait of Malacca. Some of their fears are now allayed by the construction of two new oil and gas pipelines.

Myanmar has access to the Indian Ocean and commands extensive gas reserves (the world's tenth largest). In December 2005, PetroChina entered into a covenant with Myanmar's government to purchase natural gas over a 300-year period. Based upon this agreement, CNPC (the parent company of PetroChina) signed a contract with the Daewoo International-led consortium to purchase natural gas from the Shwe gas field in A-1 offshore block in Myanmar in 2008. Thereafter, in November 2008, China and Myanmar agreed to build a $1.5 billion oil pipeline and $1.04 billion natural gas pipeline. In March 2009, China and Myanmar signed an agreement to build a natural gas pipeline, and in June 2009 another agreement to build crude oil pipeline. These two pipelines run in parallel and start near Kyaukphyu, run through Mandalay, Lashio and Muse in Myanmar before entering China at the border city of Ruili in Yunnan province. The oil pipeline (771 km long) eventually terminates in Kunming, capital of Yunnan province, The natural gas pipeline (2,806 km long) will extend further from Kunming to Guizhou and Guangxi in China. The pipeline is capable of transporting as much as 12 bcum of natural gas per year across Myanmar to China. When the oil pipeline goes operational, tankers carrying crude from distant lands of Middle East and Africa will be able to dock at Myanmar's port of Kyaukpyu and send as many as 440,000 barrels of oil a day overland to China.

All the stories narrated above paint just a small picture of the mad scramble for natural resources. If we start to write tales of many other enterprising oil and gas majors which have penetrated hundreds of countries and numerous geographies across the globe, an entire book would not suffice. Owing to constraints of space, only the important stories hinging around emerging economies have been mentioned. They explain the nature of rush and expediency of search. These vital resources have become indispensable for the better health of economies and keeping in fine fettle the engines of growth.

Avarice and Appetite for Perilous Fracking

The increasing abundance of cheap natural gas, coupled with rising demand for the fuel from China and the fallout from the Fukushima Nuclear disaster in Japan, may have set the stage for a 'Golden Age' of Gas

—Wall Street Journal, June 6, 2011

It's not unexpected that shooting massive amounts of water, sand and chemicals at high pressure into the earth to shatter shale and release natural gas might shake things up. But earthquakes aren't the worst problem with fracking

—David Suzuki, Canadian Scientist

FRACKING OR high-volume hydraulic fracturing has been generating worldwide frenzy from explorers and at the same time caused much furore among the masses. Shale is basically a sedimentary rock composed mainly of clay-size mineral grain usually of Illite, Kaolinite and Smectite. Shale may contain other clay-size mineral particles such as quartz, chert and feldspar. Other constituents might include organic particles, carbonate minerals, iron oxide minerals, sulfide minerals and traces of heavy mineral grains. Shale comes in many hues—just a small proportion of organic material or iron can significantly alter the colour of a rock, making it black, grey, red, brown or yellow. Black shale contains organic material that sometimes breaks down to form natural gas or oil. Others can be crushed and mixed with water to produce clay that can be made into a variety of useful objects. Shale is also used in production of cement in some geographies.

Shale gas has become an increasingly important source of natural gas in the United States and Canada since the start of this century, and the interest has rapidly spread to potential gas shale in the rest of the world.

Story of Shale Gas Development

After lying dormant for hundreds of millions of years, shale gas resurfaced in the first well producing natural gas from shale in 1821 in the US. Since then, industry professionals have been searching for ways to make the extraction process more efficient. Fracking, or hydraulic fracturing, was first used commercially in 1949 in the oil-and gas-rich state of Oklahoma. Horizontal drilling, came in the early 1980s and opened up access to thin layers of shale deep underground.

In the 1970s and '80s, fear of dwindling US natural gas resources galvanised research to develop methods to estimate the volume of gas in 'unconventional natural gas reservoirs' such as gas shales, tight sandstones and coal seams. This research led to the first, publicly available estimates of technically recoverable gas for gas shales in West Virginia, Ohio, and Kentucky. Because petroleum and natural gas that come from these reservoirs were challenging to extract, their recovery in the 1970s and 1980s was largely uneconomical. Thanks to improved extraction methods, the recoverable shale gas resources in the United States—nearly half of which lie in the Marcellus shale, a vast, 1-2 mile-deep rock formation stretching from southern New York through Pennsylvania, western Maryland, eastern Ohio and West Virginia—are now being tapped to provide much-needed domestic energy. Gas Research Institute's micro-seismic imaging data proved particularly useful through the 1990s up to 1997 when Mitchell Energy finally achieved economical shale gas extraction using an innovative fracturing application called slick water fracking.

US shale gas production was below 10 bcft per day during the period 2000-2008 which subsequently rose to 30 bcft/day in 2013. In 2019, it reached almost 70 bcft/day. A steep rise was seen after 2009 and continued unabated thereafter.

Massive hydraulic fracturing. Horizontal wells. 3-D mapping and seismic imaging. These, in addition to other concomitant technologies were the keys to efficient shale gas recovery as we know it today.

Global Occurrence of Shale Oil and Gas

Parameter	Countries Outside USA	In USA	Total
Shale Formations	137	-	-
Basins	95	-	-
Recoverable Shale Gas (tcft)	6634	665	7299
Shale/tight Oil (billion barrel)	287	58	345

Data source: US-based Advanced Resources International (ARI) report updated last in June 2013

Top 10 Countries with Shale Oil Resources:

Country	Shale Oil (bb)	World Ranking	Country	Shale Oil (bb)	World Ranking
Russia	75	1	Australia	18	6
US	58	2	Venezuela	13	7
China	32	3	Mexico	13	8
Argentina	27	4	Pakistan	9	9
Libya	26	5	Canada	9	10

*ARI Estimates 2013

Top 10 countries with technically recoverable Shale Gas resources:

Country	Shale Gas (tcft)	World Ranking	Country	Shale gas (tcft)	World Ranking
China	1115	1	Mexico	545	6
Argentina	802	2	Australia	437	7
Algeria	707	3	South Africa	390	8
US	665	4	Russia	285	9
Canada	573	5	Brazil	245	10

*ARI Estimates 2013

Uncertainly of Global Oil/Coal and Traction of Shale Gas

There is burgeoning interest in shale resources outside the United States. Wells drilled in shale formations in countries such as Argentina, China, Mexico, and Poland have also helped to clarify their geologic properties and productive potential. Still, many prospective shale formations, such as those underlying the large oil fields located in the Middle East and the Caspian region have not been adequately researched so far.

There has reportedly been renewed interest in shale exploration activities in a number of countries, including Algeria, Argentina, Australia, China, India, Mexico, Poland, Romania, Russia, Saudi Arabia, Turkey, Ukraine and the United Kingdom.

The Top 20 companies in shale gas exploration (in alphabetical order):

1. Anadarko Petroleum Corporation
2. Antero Resources
3. BHP Billiton
4. Cabot Oil and Gas
5. Chesapeake Energy Corporation
6. CONSOL Energy Inc.
7. Devon Energy Corporation
8. EnCana Corporation
9. EOG Resources, Inc.
10. EQT Corporation
11. ExxonMobil Corporation
12. Pioneer Natural Resources
13. Range Resources Corporation.
14. Reliance Industries Limited
15. Royal Dutch Shell
16. SM Energy Company
17. South-western Energy Company
18. Statoil
19. Talisman Energy Inc.
20. Total S.A.

Shale Gas Revolution Shapes New Geopolitical Equations

Shale gas has truly transformed the energy scene of the US. The huge surge in extraction of shale gas in the US has transformed the country's energy supply situation. In 2012, shale gas accounted for 40 per cent of total US gas production. It increased to around 25.28 tcft (trillion cubic feet) in 2019 which equals 75 per cent of total US natural gas production. This surge in domestic gas production has driven down prices and given an enormous boost to the manufacturing sector in the US.

The rest of the world is looking on with envy at the low gas prices that the US enjoys. Industrial gas prices in the US were more than four times lower than in Europe in 2012 and the wide disparity continues. Electricity prices fell by 4 per cent in the US between 2005 and 2012 while in the EU they increased by 38 per cent.

The shale gas revolution obviously brings new opportunities and choices. Given the pace of change in the shale gas and shale oil industry, it is difficult to capture all the geostrategic implications, yet a broader picture can be painted.

The greatest challenge falls upon the European Union. The EU and its member states have not really begun to discern their growing strategic vulnerability. First, it is difficult to grasp how, in the longer term, the United States can tolerate a situation in which it protects the oil flows to the EU with little or no investment by the latter in its own energy security. China may gain heavily from the shale gas revolution by reducing its energy dependency on the Middle East and reducing its supply security dependency on the US.

The shale gas revolution could also improve global food security. With the world facing an extra 2 billion mouths to feed, the availability of immense quantities of cheap gas widely dispersed across the planet is a major boon to humanity. Natural gas is the principal ingredient in artificial fertiliser. Access to cheap gas for the whole of humanity will reduce the prospect of famine and lack of nourishment worldwide.

Heightened Investment in Shale Gas

Most of the global action in shale gas arena hinges on either taking some stake in established US companies by non-US companies or US companies acquiring shale gas exploration rights outside the US.

Chinese state-owned giant CNOOC has acquired shares in American Eagle Ford and Niobrara shale oil and gas blocks. Moreover, it has made partner Chevron (US supermajor) and Royal Dutch Shell (Anglo-Dutch supermajor) explore domestic Chinese shale blocks.

Foreign companies have invested more than $26 billion in shale gas development in the US since 2008. Overall investment has crossed $133 billion for the period 2008-2012. Another Chinese company Sinochem has inked a $1.7 billion JV deal for acquiring stake in Wolfcamp shale play in west Texas.

China is abuzz with shale gas activity. It has the world's second-largest technically recoverable shale gas resources at an estimated 1,115 trillion cubic feet (tcf), after the US with 1,161 tcf.

Sinopec's Fuling shale gas field in Chongqing province is China's largest shale gas field by output with a production of 6.02 bcum in 2018. Petrochina's shale gas fields in southwest China produced 4.27 bcum of shale gas in 2018, mainly from the Changning-Weiyuan shale gas field in Sichuan. China's shale gas production has witnessed a steady rise; 4,471 bcum in 2015, 7,882 bcum in 2016 and 8,995 bcum in 2017.

China is likely to meet roughly half of its ambitious 30 bcum shale gas target in the coming years. Sinopec and Petrochina both have been driving shale projects with aggressive drilling programmes. Shale output has been growing steadily but the pace is constrained by costs and technology issues.

Shale gas production in Europe has yet to take off but Norwegian company Statoil has joined hands with US Company Chesapeake Energy to produce shale gas in Mercellus Formation in the eastern US, and has indicated interest in bringing knowledge gained in the US to European shale gas prospects.

Russian giant Gazprom has the intentions of buying US shale-gas producing company to gain expertise which it could then apply to Russian shale gas prospects. Similarly, French oil firm Total SA entered a joint venture with Chesapeake Energy (US).

Romania had recently lifted its moratorium on shale gas exploration and rights have been accorded to US giant Chevron which has started digging exploratory wells.

US ExxonMobil has leased 750,000 acres in the Lower Saxony basin of Germany, where it plans to drill at least ten shale gas wells. It is also exploring shale gas in the Mako Trough in Hungary.

Elsewhere in Eastern Europe, Poland is abuzz with activity. It has probably the largest reserves of shale gas in Europe, about 22.45 trillion cum. Of this, around 5.30 trillion cum is immediately available for extraction, tempting major global players. ConocoPhillips of US has announced its plans to explore shale gas in Poland in association with Lane Energy. Other companies such as BNK Petroleum, Talisman Energy, Marathon Oil, Chevron and ExxonMobil are equally active here and drilling exploratory wells to find the potential new gas sites.

The region with the greatest shale gas potential in Poland is the Baltic basin, where leases have already been granted to participating oil companies. The most active company here is Lane Energy, Poland's subsidiary of 3Legs Resources in a joint venture with ConocoPhilips. The results of first borings were announced on 21 June 2011. They showed immense promise.

Polish deposits are, however more difficult to access than American shale gas and entail development of different operating technologies than on the other side of the Atlantic. Hence, sufficient headway has not been made in Poland and it continues to import gas from Russia, Qatar and the US.

In Denmark, French major Total SA is active and sharing its rich experience with Denmark.V, the state-owned oil and gas company of Denmark. The Danish government has proclaimed that Denmark should be free of fossil fuels by 2050 and intends to give a major fillip to shale gas extraction.

Indian giant Reliance Industries paid $1.7 billion for a 40 per cent share in Atlas Energy's leasehold in the Marcellus shale gas play in the eastern US in 2010. This move was followed by RIL's acquisition of 45 per cent stake in Pioneer Natural Resources to gain access to Eagle Ford Shale Gas Fields. This deal was worth $1.3 billion. Another state-owned company GAIL (Gas Authority of India Ltd) acquired its maiden shale gas assets in the US through its wholly owned US subsidiary GAIL Global (USA) Inc for $95 million in September 2011. By this deal, GAIL acquired a 20 per cent interest in Carrizo Oil's Eagle Ford Shale acreage. The Houston-based firm Carrizo Oil has a joint venture with RIL too for Marcellus shale acreage in northern and central Pennsylvania. This momentous deal envisaged total investment worth around $300 million over a period of five years. Thus, both Reliance and GAIL from India have emerged pioneers in overseas shale gas assets acquisition.

In the African continent, the action is concentrated in South Africa, which has a major sedimentary basin, Karoo Basin that contains thick organic-rich shales. The Karoo is estimated to have technically recoverable resource of around 500 trillion cubic feet. Falcon Oil and Gas Ltd of Ireland was an early entrant into the shale gas play in South Africa, obtaining an area of 30,000 sq km Technical Cooperation Permit (TCP) along the southern edge of the Karoo Basin. Royal Dutch Shell obtained a still larger 185,000 sq km TCP surrounding the Falcon area, while Sunset Energy holds about 4,600 sq km TCP to the west of Falcon. The Sasol, Chesapeake, Statoil are also active here.

Canada has a number of prospective shale gas sites in various stages of exploration and exploitation in British Columbia, Ontario, Quebec, New Brunswick, Nova Scotia, Alberta and Saskatchewan. Encana and EOG Resources are developing the area known as Horn River, believed to possess huge shale gas deposits.

Shell Global has reached many geographies and is producing shale gas in North America: Groundbirch and Deep Basin (Canada), Pinedale (Wyoming); Haynesville (Louisiana); Eagle Ford (new Texas acreage) and Marcellus Shale. In western Canada, Shell acquired the Duvernay Natural Gas Company in 2008. They have also penetrated South Africa, China, Eastern Ukraine, Australia and Germany.

Global major MC (Mitsubishi Corporation) and Korea based KOGAS (Korea Gas Corporation) entered into partnership in 2011 to develop natural gas project centred on shale gas in the Cordova Embayment of north-eastern British Columbia, Canada. Other Japanese companies like JOGMEC (Gas and Metals National Corporation), Chubu Electric, Tokyo Gas and Osaka Gas each have also acquired stake in this tripartite project involving Canada, Korea and Japan.

Malaysian oil and gas company Petronas through its overseas arm PICL has established a strategic partnership with Canada-based Progress Energy Resources Corporation to develop its shale gas assets into LNG in north-eastern British Columbia. The venture is meant to monetise these assets by liquefying shale gas into LNG before marketing it to the lucrative East Asian market.

Cuadrilla, a British oil and gas company, ended its long-running legal battle in October 2018 allowing it to restart fracking in the UK. In 2011, all fracking in the UK was suspended after evidence suggested it had caused minor earthquakes in the north-west of England. The UK government's support for fracking however, has been denounced by many, including climate scientist James Hansen, who opines that fracking would undermine the UK government's efforts to meet climate obligations.

Indian Companies in Shale Gas Race

Shale gas exploration in India is currently in incipient stage and full-scale commercial production has not commenced yet. Apart from RIL and GAIL which have gone for overseas shale gas projects, ONGC and OIL were engaged by the government for domestic shale gas exploration. US Energy Information Administration (USEIA) estimates put India's total recoverable shale gas potential around 584 tcft. India announced its New Shale Gas Policy, according to which it allowed ONGC and

OIL the right to explore shale gas and oil in their nomination blocks. While ONGC would take up 175 blocks, Oil India will do it in 15 blocks in three assessment phases. ONGC will get 50 blocks in the first phase, 75 blocks in the second and 50 blocks in the third phase. Oil India would explore five blocks each in all three phases.

ONGC first struck shale gas in a pilot project at Ichhapur in Burdwan, West Bengal. The drilling of Jambusar-55 (Gujarat), the first well under the pilot programme, started on 27 October 2013. ONGC had also signed a memorandum of understanding with the ConocoPhillips, US and the two undertook joint studies of the four Indian basins: Krishna Godavari, Cambay, Cauvery and Damodar.

In November 2019, ONGC wound up its shale exploration programme mid-way after spending six years and hundreds of crores of rupees, concluding that India might not have enough commercially extractable shale gas reserve.

The US has permian shale while India has much younger, tertiary shale. The US shale rocks are brittle and so rather easy to hydro-frack, while those in India are elastic with more clay content, resulting in little yield from fracking.

Past estimates of India's shale reserves also vary widely from ONGC's 187 tcf in five basins to USEIA's 584 tcf of shale gas and 87 billion barrels of shale oil in four basins, and oilfield services provider Schlumberger's 300-2,100 tcf gas.

A new official assessment in alliance with an international agency may bring a clearer picture on shale reserves. A thrust on research and development and strong incentives to services and technology companies to set up base in India will encourage creation of advanced knowledge and help further exploration.

Emerging New Technologies

Shale gas is held in rock pores up to 20,000 times narrower than a human hair. Often the gas will not flow freely into a well, or it flows at a much slower rate than in normal gas reservoirs. The amount of gas that would be recovered from each well would be low but the overall volume of available gas in the reservoir can be much higher than conventional gas reservoirs. Technology is needed to produce it safely, economically and in a way that helps protect the environment. Let's have a peek into new promising technologies of Super Fracking, RapidFrack, HiWay and Fracking through Fibre Optical Cables (FOC),

Fracking through FOC

At present, fracking each well typically takes at least 15 stages of mixture-firing. To improve on the existing methods, Halliburton Company from New York and others are testing military grade fibre optic cables that are used in US submarines. These fibre optic lines record sound and temperature along their entire length. With steel-encased lines clamped between fracking wells and rock, drillers can record sounds that signal the perfect frack. Efforts are on to convert the sounds to a graph, showing how thoroughly the rock hiding the fuel has been fractured. This would enable to make the perfect frack with great precision every time.

RapidFrack

RapidFrack is being touted as 'the frack of the future.' This system uses a metering process that allows access to multiple fracture points to enhance production and reduce the volume of water used during the fracking process.

HiWay

Schlumberger has developed an advanced fracking technology called HiWay after six years of research. HiWay injects specially treated material into fractures to widen pathways for the flow of oil and gas, also reducing water consumption.

Resistance to Fracking

Stiff resistance and violent demonstrations against fracking are reported at regular interval across the globe. Even in the US and Canada, periodic protests are observed at the existing or new fracking sites. Many groups have sprung up seriously campaigning against fracking.

Tapping of shale gas by Chevron in the European Union's second-poorest state of Romania triggered nationwide protests in September 2013, throwing together local communities, environmentalists, civic rights groups and the clergy. USEIA estimates put Romanian recoverable shale gas reserves around 51 tcf. Chevron had to suspend its exploration plans in Pungesti in Vaslui County, 340 km north-east of the capital Bucharest after locals blocked access to the site.

Anti-fracking protests descended into serious violence in New Brunswick, Canada in October 2013. Canadian protesters set cars on fire and threatened officers mounted on horseback. Activists and local aboriginals held demonstrations for an entire week to protest shale gas exploration in the region.

Despite a quick embrace by the resource-hungry states, a large number of countries are still vehemently opposed to fracking. Let's cast a quick glance over the parts of the globe where the practice of fracking is deeply abhorred.

France was the first country to impose ban on fracking

in 2011 followed by Bulgaria in 2012. Other prominent countries from Europe with fracking ban are Germany and Ireland. The US state of Vermont became the first state in the US to ban fracking on 17 May 2012. A large number of people are opposed to shale gas exploration in Canada and Quebec and British Columbia have promulgated laws banning the practice. South Africa and the UK did impose bans which they later lifted. Globally, around 420 measures have been passed pertaining to ban on fracking in various cities/counties/states.

Environmental Burden and Health Hazards of Fracking

Shale gas extraction has largely lowered the cost of electricity, industrial feedstock, residential and commercial energy and even spurred the use of natural gas as a transportation fuel. Nevertheless, shale gas development is extremely controversial and a highly debated issue in that the potential health and environmental risks related to drilling and production activities are still poorly or inadequately understood. Washington-based research organisation Resources for the Future (RFF) found and analysed 'Risk Matrix' which included about 264 'Risk Pathways' associated with shale gas exploration and production activities.

Shale gas extraction raises environmental concerns in relation to the following:

- Carbon dioxide (CO_2) and methane (CH_4) emissions, particularly the potential for increased fugitive CH_4 emissions during drilling compared with drilling for conventional gas
- The volumes of water and the chemicals used in fracking and their subsequent disposal
- The possible risk of contaminating aquifers and groundwater
- Competing land-use requirements in densely populated areas

- Physical effects of fracking in the form of increased seismic activity and induction of earthquake
- The potential impact on surface water from using wastewater for road de-icing or dust suppression
- Seismic vibrations caused by deep underground injection of flowback and produced water
- Environmental burdens of air pollutants, intrusion of saline water into aquifers, noise pollution and road congestion as a corollary to shale gas activities
- Associated problems of on-site pit and pond storage of flowback liquids.

Some studies conducted in recent years have brought out health effects of fracking. One study carried out in rural Colorado covering around 1.25 lakh births during 1996-2009 points out an association between density and proximity of natural gas wells within a 10-mile radius of maternal residence and prevalence of congenital heart defects and possibly neural tube defects. However, greater specificity in exposure estimates is needed to further explore these associations. Another study reveals that residents living within half a mile from wells are at greater risk for health effects from natural gas development than are residents living over half a mile away from wells. Sub-chronic exposure to air pollutants during well completion activities present the greatest potential for health effects. Cumulative cancer risks were 10 in a million and 6 in a million for residents living within half a mile and greater than half a mile from wells, respectively, with benzene as the major contributor to the risk.

Around 1,000 products containing over 600 chemicals are

used during various stages of fracking. Some of these chemicals may affect the skin, eyes, other sensory organs, and the respiratory and gastrointestinal systems. Similarly a number of these chemicals could affect the brain/nervous, endocrine, immune and cardiovascular systems and the kidneys. At least one-fourth of the deployed chemicals are potential carcinogens and could cause cancer and mutations. As fracking is a rather new phenomenon, long-term data and studies are further required to convincingly corroborate the above findings. Nonetheless, we should not take these findings lightly.

Blind Rush and the Road Ahead

Sudden spurt in global activities and undue haste to have access to this newly discovered natural resource has resulted in waves surging across the continents with the US at the epicentre. Now the cardinal question arises: Is the world decisively ready for full-scale fracking disregarding pros and cons? That's a moot issue for heated debates. No doubt, shale gas has intrinsic benefits. If extracted prudently, it generates numerous jobs and big bucks in revenue. It could arguably lower the global greenhouse gas emissions as well. Natural gas emits less carbon dioxide per unit burned than coal and gasoline, but in its unburned state, it is itself a more potent greenhouse gas than carbon dioxide. At least three dozen countries/states around the globe have put a moratorium on fracking and in many states the bills seeking ban are pending or under consideration.

Any new development often brings social costs as an adjunct. Take the case of Pennsylvania, where natural gas boom has brought thousands of new gas wells, a number of transient workers and a host of associated social problems.

Traffic accidents, civic disturbances and public health problems in rural Pennsylvania counties have soared since the shale rush began in 2005, diminishing the quality of life for the residents. These social and public health costs escalated more in rural counties with the new shale gas wells than in rural counties without shale gas drilling. These negative social impacts were especially pronounced in the counties with the preponderance of shale gas wells.

The fracking boom often engenders vehicular crowding especially heavy trucks swarming rural areas and gives rise to out-of-state workers flooding small towns, often overwhelming local housing, policing and public health capacities. The influx of transient workers with disposable income and little to do in their off hours is a recipe for trouble. Alcohol-related crimes, traffic accidents, emergency room visits and sexually transmitted infection all are tremendously augmented.

Till date, much of the discussion about fracking has focused on the obvious environmental risks, while the social costs have been largely ignored. They need to be judiciously evaluated before blindly giving thumbs up to hydraulic fracturing.

Lessons Learnt

The great success story of the US had not been emulated by any country until 2019. Commercial production outside the US is restricted to Canada, Argentina and China only. However, China has been desperately trying to become another great player in the domain of shale gas extraction with full state support and backing.

The main conclusion that could be drawn from various stories beyond North America is that the United States model is unlikely to be directly replicable in other countries. Production techniques need to be adapted to local geological conditions before commercial production starts. Countries will therefore have to pass through an experimental stage that may appear long and costly. Finally, general acceptance by local populations may prove pivotal.

Across Europe, high population density, the absence of a social licence to operate, population's antipathy to hydrocarbon activities beside habitations pose serious hindrances in exploiting shale gas. Apart from this, the nature of local mineral rights which accord ownership of underground resources to the state, have not prompted populations to welcome the development of shale gas activity, evident in large-scale demonstrations across Europe. As a result, several countries have successively put a moratorium on shale gas activities or hydraulic fracturing—with the exception of Poland.

The unprecedented and rapid expansion of shale gas production in the US has also led to surfacing of major concerns over the potential negative impacts of hydraulic fracturing on the environment, especially with regard to water pollution, large levels of water withdrawals and induced seismicity.

Water-related issues form cardinal criticism points. Among potential issues are the contamination of surface and groundwater by fracturing fluid, flow-back waste water, fugitive CH_4, as well as the use of the large volumes of water necessary to perform fracturing operations. Water is of paramount importance to life on Earth and this is all the more important in areas where water availability is already an issue or in regions where new activities would compete with existing activities consuming large amounts of water otherwise available for irrigation and drinking.

Consequent upon Paris Accord 2015 (effective since 4 November 2016), international commitments pledged by countries aimed at curbing global warming, the share of renewable energy (solar/wind) in the global energy mix is expected to expand and may account for 19 per cent of the global energy mix by 2035. With this in mind, a key issue relating to the current trends in the shale gas sector is to evaluate the role that natural gas is expected to play in the future energy landscape. Natural gas is pointed out as a bridge fuel by the Intergovernmental Panel on Climate Change (IPCC), when it is used as a source of energy that emits less CO_2.

Although it is less carbon intensive than other fossil fuel (oil and coal) yet natural gas also emits CO_2 into the atmosphere and contributes to global warming. Moreover, the issue of fugitive CH_4 emissions (GWP potential 34) must be particularly taken into consideration and mitigation strategies applied. In addition, irrespective of the level of future prices of natural gas, its use should contribute to fostering a smooth transition only from the current economic model, mainly based on fossil fuels, to achieving a low-carbon growth trajectory.

Still Crazy for Nuclear Energy!

To meet our growing energy needs and prevent the worst consequences of climate change, we need to increase our supply of nuclear power

—Former President US Barack Obama

I foolishly once believed the myth that nuclear energy is clean and safe. That myth has completely broken down. Restarting nuclear reactors while we still have no place to dispose nuclear waste is a criminal act toward further generations

—Morihiro Hosokawa; 79th PM of Japan

OKUMA, A PLACID town located near the beautiful Pacific coast, which once boasted of the prestigious Fukushima Daiichi Nuclear Power Plant (NPP), has now lost much of its sheen. Roads leading to the plant wear a quiet deserted look. Hardly a man is seen for several kilometres from the plant except the workers around the reactors combating the unceasing leakage, risking fatal radiations. Following the Tohoku earthquake and resulting tsunami on 11 March 2011 which crippled the NPP housing six reactors, strange things are occurring and being reported in and around the area. Ramifications are being felt as far as the US coast. Some animals have developed white spots all around the body, a possible sign of early cancerous lesions. Many people in Japan and the US coastal area reported metallic taste in the mouth lasting for several days, ceasing and re-appearing—a tell-tale sign of radiation exposure. Yellow rain was reported in Tokyo and surrounding areas on 23 March 2011 that also happened after Chernobyl. A mutant rabbit was born in Namie City, just outside the 30-km exclusion zone surrounding the Fukushima Daiichi NPP. The rabbit has no ears and bears Albinism, which is actually caused by a mutation in several genes. Just imagine if these things start replicating in human beings!

Fukushima continued to spew massive amounts of radiation into the groundwater, into the ocean, and into the air several years after the reactor meltdown. Tepco dumped 400 tonnes a day, plus up to 10,000 tonnes of high-level liquid radioactive waste water into the ocean after the accident.

Radioactive material cesium has been found in a surprising place—beneath the coastal beach sands close to brackish water tables up to sixty miles away from the meltdown site. Cesium-137, which exists as residual background radiation leftover from nuclear testing in the mid-1900s (with a half-life of 30 years), is still being detected along the entire west coast at levels higher than would otherwise be present, though still well below government-established safety limits.

It is believed that where currents washed highly contaminated water onshore, cesium stuck to sand particles in brackish water. Therefore, the accumulated cesium embedded under beaches along Japan's coast is being slowly released over time as waves saturate the sand. Scientists estimate that the amount of cesium being released in this manner is equal to that of the ongoing releases and runoff from the disaster site itself, and from outflow of rivers continuing to carry cesium from land to the sea.

The Fukushima crisis is far from over, however. Millions of tonnes of contaminated, radioactive water (used to keep the molten reactors and spent fuel rods cool to prevent further radioactive releases, and from pumped groundwater that continues to seep into the reactors' basements) is stored on-site in more than 900 tanks. Despite efforts to prevent seepage using 'ice walls' (man-made permafrost), and special processing technology intended to clean radionuclide from the water, Tepco was forced to admit in 2018 that these efforts had failed.

While more water continues to become contaminated every day, the only known plan to deal with it is to release it all into the ocean. This is of great concern because of the dangerous nature of these particular isotopes. Strontium-90, for example, has a half-life of 28.8 years and mimics calcium once ingested. If ingested by fish, or further up the food chain by people, it becomes stored in the bones, which greatly increases the risk of cancer. If this stored water is released into the ocean, fish that migrate long distances could potentially carry these isotopes across the Pacific.

As quoted in the *South China Morning Post*, Shaun Burnie, a nuclear specialist with Germany's Greenpeace stated, 'The Japanese government and Tepco set an objective of "solving" the radioactive water crisis by 2020—that was never credible. The reality is that there is no end to the water crisis at Fukushima, a crisis compounded by poor decision-making by both Tepco and the government. Discharging into the Pacific is the worst option and must be ruled out.' It remains to be seen what will ultimately become of the situation.

The description above depicts the horrendous picture of the nuclear fallout that happened in Fukushima prefecture in Japan.

The idea of an atom began with the Greek philosopher Democritus, who propounded all matter consisted of tiny particles. He called them 'atomos' the Greek word for 'indivisible'. He couldn't prove they existed but centuries later other scientists did. That discovery heralded the evolution of nuclear power in later times, simultaneously hogging controversy and a serious debate in recent times.

Now, let's review the evolution of nuclear energy over the past 5-6 decades or so. Nuclear technology was first evolved in the 1940s during World War II. Initial research was mainly focused on producing bombs by splitting the atoms of particular isotopes of either uranium or plutonium. During the 1950s, however, attention shifted to the peaceful purposes of nuclear fission, notably for power generation. Thus, the first commercial nuclear power stations started operation in the 1950s.

There are now over 430 commercial nuclear power reactors operational in 31 countries, with over 370,000 mwe of total capacity. Many more reactors are under construction. They provide over 11 per cent of the world's electricity as continuous, reliable base-load power; without carbon dioxide emissions. Fifty-six countries operate a total of about 240 research reactors and a further 180 nuclear reactors energise around 150 ships and submarines. Today, the world produces as much electricity from nuclear energy as it did from all sources combined in 1960.

Nuclear power is considered to be emission free, producing minimum pollution, yet when accidents occur either due to human negligence or natural disaster, the consequences are often catastrophic. The long-term effects of nuclear disasters can often spread over thousands of years. It is believed that Chernobyl is not likely to be fully inhabited for at least another 20 millennia. At Fukushima, it would take several decades to clean up and arrest seeping water containing highly radioactive substances. Let's review the previous prominent nuclear disasters that occurred in different geographies.

1. Chernobyl, Ukraine (1986)

The Chernobyl nuclear disaster in Ukraine is widely recognised to have been the worst power plant accident in the annals of nuclear disasters. This disaster is accorded Level 7 event on the International Nuclear Event Scale (INES). The battle to contain the contamination and avert this mammoth catastrophe involved some 500,000 workers and cost an estimated 18 billion roubles. Death figures are still disputed but certainly reckonable. Long-term effects such as cancer and deformities are still being accounted for.

2. Fukushima, Japan (2011)

The Fukushima nuclear fallout resulted in a series of equipment failures, nuclear meltdowns and releases of radioactive materials in the wake of Tohoku tsunami on 11 March 2011. It is only the second disaster along with Chernobyl to measure at Level 7 on the INES. Till date, radioactive and highly contaminated water continues to spill from the leaked reactor, seriously polluting the ground water. It has been a nightmare for Tepco authorities to contain the continuously spilling water, though several methods have been experimented with. Large swathes of land now bear soil that has been rendered useless as well as dangerous owing to fatal levels of radiation contamination. A herculean task has been undertaken to remove this highly contaminated soil and bury it safely.

3. Kyshtym, Russia (1957)

The Kyshtym nuclear disaster was a radiation contamination incident that occurred on 29 September 1957 at Mayak, a nuclear fuel reprocessing plant in the Soviet Union. The event at Level 6 occurred in the town of Ozyorsk, a closed city built around the Mayak plant. Since Ozyorsk/Mayak were not marked on maps, hence the disaster was named after Kyshtym, the nearest known town.

4. Three Mile Island Disaster, Pennsylvania, US (1979)

Two nuclear reactors suffered meltdown on 28 March 1979. A Level 5 disaster, it is often reckoned the worst disaster in commercial NPP history. Small amounts of radioactive gases and radioactive iodine were released into the environment.

5. Windscale Fire, UK (1957)

This Level 5 disaster on 10 October 1957 is considered the worst nuclear disaster in Britain's history. The accident occurred when the core reactor of Unit – 1 caught fire, releasing substantial amounts of radioactive contamination into the surrounding area. As many as 240 cancer cases have since been linked to the fire. All the milk within a 500-km radius was diluted and destroyed for about a month.

Nuclear accidents keep happening throughout the world, at least 100 having been reported so far. At least 56 accidents occurred at nuclear reactors in the US alone, followed by Japan and France registering 12 each. Canada, India, Germany and the UK have also witnessed nuclear accidents. Other noteworthy disasters of Level-4 include Tokaimura, Japan 1999, Buenos Aires, Argentina 1983, and Saint-Laurent, France 1969. They have caused significant damage to human life and property. It is estimated that nuclear accidents resulted in a loss of over $20 billion in the US alone during the period 1952-2009. It becomes evident this cheap power doesn't come so cheap. However, staunch nuclear proponents argue car accidents also keep happening rather frequently around the world, that does not mean we should abandon car use! Well, we shall peek into this aspect a little later. Before that, let's see how inevitable nuclear power has become across the globe.

Growing Global Dependence on Nuclear Power

What's so tempting about nuclear power? Well, NPPs are by far the lowest-cost producers of base-load electricity. By burning one kilogram, coal can illuminate a 100-watt light bulb for about four days, natural gas can do this for about six days, and uranium for 140 years. In other words, a single uranium fuel pellet the size of a pencil eraser contains the same amount of energy as 17,000 cubic feet of natural gas, 1,780 pounds of coal or 149 gallons of oil. Its fuel is on site (not depending on continuous delivery), it is dispatchable on demand, it has fairly quick ramp-up, it contributes to clean air and low-CO_2 objectives, it gives good voltage support for grid stability. These attributes are mostly not monetised in merchant markets, but have great value which is increasingly recognised where dependence on intermittent sources has grown. Moreover, this power comes with zero emissions. Hence the huge traction and immense craving for this energy.

Although only eight countries (US, UK, Russia, France, China, India, Pakistan and North Korea) are known to possess nuclear weapons capability as of now, at least 56 countries operate around 240 civil research reactors, over one-third of these being in developing countries alone. Thirty-one countries now host over 430 commercial nuclear power reactors with a total installed capacity of over 370,000 mwe . This is more than three times the total generating capacity of France or Germany put together from all sources. Around 70 more nuclear power reactors are under construction, equivalent to 20 per cent of existing capacity, while over 160 are firmly planned that tantamount to half of the present capacity.

Sixteen countries depend on nuclear power for at least a quarter of their electricity. France gets around three-quarters of its power from nuclear energy while Belgium, Czech Republic, Hungary, Slovakia, Sweden, Switzerland, Slovenia and Ukraine get one-third or more. South Korea, Bulgaria and Finland normally get more than 30 per cent of their power from nuclear energy, while in the US, UK, Spain and Russia almost one-fifth is nuclear. Japan is used to relying on nuclear power for more than one quarter of its electricity and is expected to return to that level. Among countries which do not host NPPs, Italy and Denmark get almost 10 per cent of their power from nuclear plants.

Nuclear electricity constitutes nearly 12 per cent of total world production. Graphic 8.1 illustrates the countries and their percentage share of nuclear electricity in total domestic electricity production.

Graphic 8.1: Percentage of Nuclear Electricity in Total Domestic Electricity Production

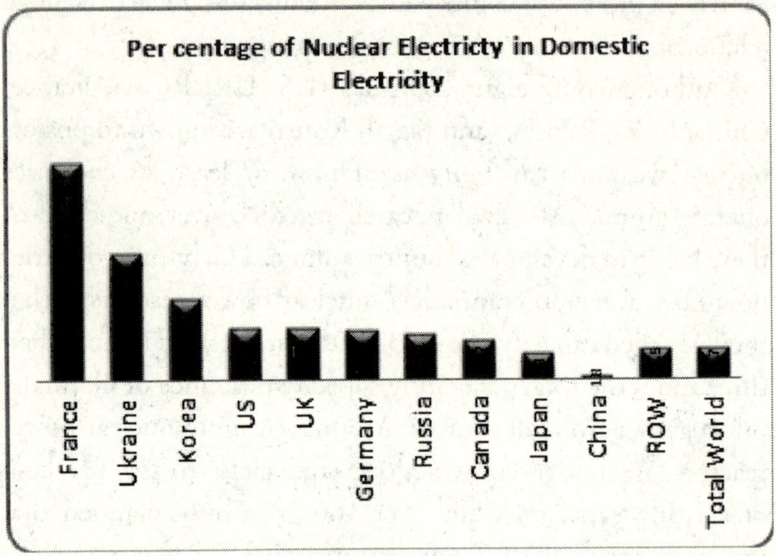

*Data Source: International Atomic Energy Agency

** ROW-Rest of the World

Old Obsession Continues

Most NPPs were originally designed with a nominal design life of around 25-40 years, but fresh engineering assessments of many plants have established that many can operate longer. In the US alone, over 70 reactors have been granted licence renewals which extend their operating lives to 60 years. Many more operators are expected to apply for similar extensions. In France, there are rolling ten-year reviews of reactors. In 2009, the Nuclear Safety Authority (ASN) approved EdF's safety case for 40-year operation of the 900 mwe units, based on generic assessment of the 34 reactors. The Russian government is also extending the operating lives of most of the country's reactors from their original 30 years, for 15 years, or for 25 years in the case of the newer VVER-1000 units, with significant upgrades.

World's Top Nuclear Electricity Producers

The US tops among all nuclear producers registering 821 twh (terawatt hour, a measure of electric energy) in 2011. France trails behind the US with 442 twh production. It is interesting to note that more than four-fifths of total world's nuclear electricity production comes just from top ten countries. Rest of the world put together produces only less than one-fifth. Moreover, just two countries (US and France) contribute nearly half of the total nuclear production. Top producers have been shown in Graphic 8.2 along with their percentage share.

If we analyse the data for 2011 and 2018, it is amply evident that western countries like France, Germany, the UK and the US have lowered their production of nuclear energy. In Asia, South Korea has also lowered its nuclear production but China has significantly increased it from 86 twh to 277 twh, amounting to three-fold increase in just seven years!

Graphic8.2: Top Nuclear Electricity Producers in 2011

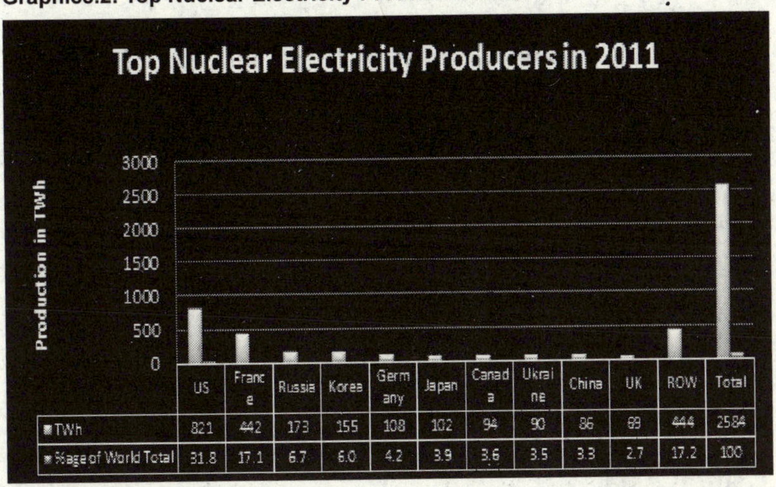

*Data Source: International Atomic Energy Agency

** ROW-Rest of the World

*** TWh-Terawatt hour

Graphic 8.3: Top Nuclear Power Producers in 2018

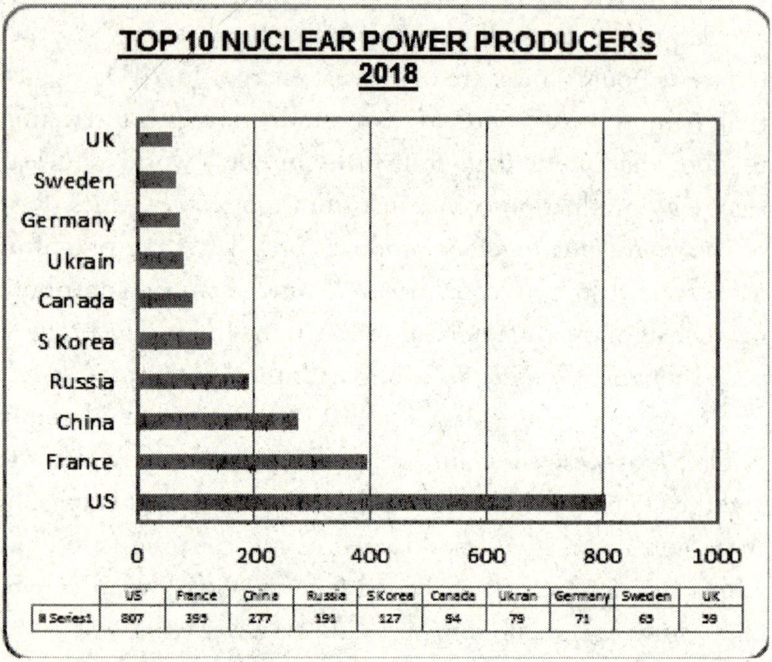

*Data Source: US Energy Info. Adm./IAEA 2019

OVERVIEW OF NUCLEAR POWER DEVELOPMENT

Here is the brief description of various countries' nuclear development plans.

Africa

South Africa is committed to plans for further conventional nuclear power reactors. Nigeria has sought the support of the International Atomic Energy Agency to develop plans for two 1,000 mwe reactors.

Central Asia

Kazakhstan with its abundance of uranium is working closely

with Russia in planning development of small new reactors for its own use and export.

Europe

Finland and France are both expanding their fleets of NPPs with the 1,650 mwe EPR from Areva, two of which are also being built in China. Several countries in Eastern Europe are currently constructing or have firm plans to build new NPPs (Bulgaria, Czech Republic, Hungary, Romania, Slovakia, Slovenia and Turkey).

A UK government energy paper in mid-2006 endorsed the replacement of the country's ageing fleet of nuclear reactors with new nuclear build, and four 1,600 mwe French units are planned for operation by 2023. The government aims to have 16 gwe of new nuclear capacity operating by 2030.

Sweden has abandoned its plans to prematurely decommission its nuclear power and is now investing heavily in life extensions and updates. Hungary, Slovakia and Spain are all implementing or planning for life extensions on existing plants. Germany agreed to extend the operating lives of its nuclear plants, reversing an earlier intention to shut them down but reversed policy following the Fukushima accident.

Poland is developing a 6,000 mwe nuclear programme. Estonia and Latvia are involved in a joint project with established nuclear power producer Lithuania. Belarus has started construction of its first Russian reactor and a second is due to follow.

India

India's aim is to have 14.5 gwe of nuclear capacities on line by

2020 as part of its national energy policy. Its reactors include light and heavy water reactors as well as fast reactors. Seven power reactors are under construction or almost so, of both indigenous and foreign design, including a 500 mwe prototype fast breeder reactor. This will take India's ambitious thorium programme to Stage 2, and set the scene for eventual utilisation of the country's abundant thorium-to-fuel reactors.

Though India has ambitious plans to catapult its nuclear capacity to 63 gwe, yet total capacity is likely to be only about 22.5 gwe by the year 2031.

Constraints are owing to fundamental incompatibility between India's Civil Liability Law and international conventions limiting foreign technology. Apart from this, India was largely excluded for trade in nuclear plant and material for nearly 34 years because of the Nuclear Non-Proliferation Treaty (NPT) which hampered development of civil nuclear energy until 2009. Its current capacity is 5.4 gwe.

Middle East

The United Arab Emirates (UAE) is building the first two of four 1,450 mwe South Korean reactors at a cost of over $20 billion and is collaborating closely with IAEA and experienced international firms. Iran's first power reactor is in operation, and more are planned. Saudi Arabia, Jordan and Egypt are also moving towards employing nuclear energy for power and desalination.

People's Republic of China

To increase its nuclear generating capacity to 58 gwe, China commenced operation of 17 new nuclear power reactors over 2002-2013. Thirty reactors are under construction. These include the world's first four Westinghouse AP1000 units and a demonstration high-temperature gas-cooled reactor plant. Many more are planned with construction due to start. China is commencing export marketing of a largely indigenous reactor design. R&D on nuclear reactor technology in China is unequalled.

China has now 45 nuclear power reactors in operation and 12 more are under construction as of February 2020. Its current installed nuclear power generating capacity is 33.6 gwe.

Russian Federation

Russian government has also set a target to increase its nuclear capacity to 30.5 gwe using world-class light water reactors. Construction of a large fast breeder unit is nearly complete, and development proceeds on others, aiming for significant exports. Russia has one floating n-pwer plant and plans to build seven more by the next decade. Russia is also active in building, financing and technology transfer of new NPPs in several countries.

South America

Argentina and Brazil both have commercial nuclear reactors generating electricity, and additional reactors are under construction. Chile has a research reactor in operation and has the infrastructure and intention to build commercial reactors.

Asia

Vietnam intends to have its first NPP operating about 2023 with Russian help and a second soon after with Japanese input. Indonesia and Thailand are planning nuclear power programmes.

South Korea has plans or placed orders for 12 new nuclear power reactors. It is also involved in intense research on future reactor designs.

Bangladesh has approved a Russian proposal to build its first NPP. Pakistan with Chinese help is building three small reactors and preparing to build two large ones near Karachi.

United States

In the US, there are five reactors under construction, four of them new AP1000 designs. Two plants are under construction and would start commercial production in 2021 and 2022 respectively. Construction of two plants is halted. One of the reasons for the hiatus in new build in the US to date has been the extremely successful evolution in maintenance strategies. Changes have increased utilisation of US NPPs, with the increased output corresponding to nineteen new 1,000-mw plants being built.

North Korea

Despite repeated rebukes and UN sanctions, North Korea continues with nuclear tests time and again. Its main nuclear facility is believed to be Yongbyon site. It first conducted nuclear tests in 2006, again in 2009 and yet again in 2013. Analysts believe the first two tests used plutonium. The country is now widely believed to possess enough weapons-grade plutonium for at least six bombs. Whether it used plutonium or uranium for the 2013 test is unclear. North Korea's nuclear actions often provoke the international community and have become a cause of concern among its neighbours, especially South Korea.

North Korea made certain claims about its capabilities in

the wake of its 2013 test. Firstly, it claimed it had made a device small enough to fit a nuclear warhead onto a missile. If true, this is likely to be of serious concern to the US and North Korea's neighbours. It also claimed its latest test had a much greater yield than the plutonium devices it detonated in 2006 and 2009.

Several rounds of negotiations have taken place between the North Korea, the US, Russia, China, Japan and South Korea aimed at persuading Pyongyang to give up its nuclear ambitions. None of this has ultimately deterred North Korea.

North Korea's proclamation of a 'high-level' test could have been code for the use of highly enriched uranium (HEU) instead of plutonium. Although both represent roughly the same level of threat, a uranium bomb would signify a huge technological achievement and a grave international threat, given that it is in whimsical hands.

Iran

Whether Iran is hurtling toward nuclear weapon capabilities is one of the most contentious issues challenging the West, including the United States and Israel. Cancelled after the 1979 revolution, Iran's nuclear programme was revived in the closing phases of the war with Iraq. Teheran wanted to guard against a future surprise analogous to Iraq's repeated use of chemical weapons.

Iran envisages an energy programme that encompasses nearly a dozen reactors generating around 24,000 mw and several enrichment plants. It is also building a heavy-water plant at Arak, a source of potential proliferation concern.

Teheran repeatedly proclaims its quest to acquire peaceful nuclear energy to serve a population that has doubled since the 1979 revolution and explicitly denies any interest in developing a nuclear weapon. Sometimes though, it boasts of its growing ability to enrich uranium, a capability that can be used to generate power or for a weapons programme. Technically, Iran does not yet need to enrich, since Russia is providing the fuel for the new Bushehr reactor. Teheran argues it has the right to enrich uranium as a signatory to the NPT. It also intends to build additional reactors and says it does not want to be dependent on foreign powers for fuel.

Since 2002, international suspicions about Iran's long-term intentions have deepened because of revelations by other governments or Iranian exile groups that it has built secret facilities that could be used for a weapons programme. Bushehr 1,000-mw light water reactor was built through Russian collaboration in 2011 and took 15 years to complete. The deal stipulates that fuel is provided by Russia and the spent fuel rods will return to Russia. Iran is also working on a new 360 mw Darkhovin NPP. As of 2015, Iran's nuclear programme has cost $100 billion in the form of foreign direct investment (FDI) and oil revenue loss.

In July 2019, IAEA stated Iran had breached the Joint Comprehensive Plan of Action (JCPOA. In January 2020, following the killing of Commander Qasem Soleimani, Iran stated it would no longer abide by the JCPOA. Latest information by the IAEA (confirmed in March 2020) reveals that Iran has nearly tripled its stockpile of enriched uranium since November 2019 and that is quite dangerous to world peace and stability in the Middle East.

Surge in Uranium Mining

Most uses of uranium exploit the nuclear properties of this

mineral. Uranium-235 is the only naturally occurring fissile isotope. It can be transmuted in a nuclear reactor into fissile plutonium-239. Uranium is also used in nuclear weapons but it also has some other uses. Uranium glass produces orange and red to lemon hues. It was used for shading and tinting in early photography. Its radioactive properties were discovered back in 1896. Enrico Fermi and others conducted research starting 1934 which led to the use of uranium as fuel in the nuclear industry and in the bomb Little Boy dropped in World War II on the Japanese.

In 2011, the world consumed about 178 mt of uranium including buffer stock. Mines only produced about 138 mt. The uranium production industry witnessed the consolidation phase by takeovers, mergers and closures during the 1990s. However, this has diversified in recent years with Kazakhstan's diverse ownership structure. There was a smart recovery of uranium prices around 2003; there has been a lot of activity in preparing to open new mines in many countries. The World Nuclear Association (WNA) predicts world uranium demand at about 72,680 tU in 2015, and most of this will need to come directly from mines (in 2010, 22 per cent came from secondary sources and this shrunk to 14 per cent in 2012). Some mines like Khiada (Russia), Four Mile (Australia), Cigar Lake (Canada) are expected to reach substantial production, Husab (Namibia) and Mkuju River (Tanzania) are expected to yield full potential by 2015 and 2016 respectively. Another mine Imouraren in Niger is expected to be operational in coming years.

Estimated future production from existing mines plus new projects such as these are however not sufficient to meet the demand requirements of 2030. A recent report by the WNA projects huge mismatch in demand-production scenario. It estimates demand 119,000 tU in upper scenario and 97,450 tU in reference scenario by 2030 whereas production around 60,000 tU in upper scenario and 53,000 tU in reference scenario.

As of now, around 65 per cent of the world's production of uranium is coming from mines in Kazakhstan, Canada and Australia. An increasing proportion of uranium, now 46 per cent, is produced by *in situ* leaching (ISL). After a decade of falling mine production up to 1993, output of uranium has generally risen since then and now meets over 90 per cent of demand for power generation. Kazakhstan produces the largest share of uranium from mines (38 per cent of world supply from mines in 2013), followed by Canada (16 per cent) and Australia (11 per cent).

Mining methods have also been changing. In 1990, 55 per cent of world production came from underground mines, but this shrank dramatically in 1999, with 33 per cent then. From 2000, the new Canadian mines increased it again, and with Olympic Dam it is now 37 per cent. ISL mining has been steadily increasing its share of the total, mainly due to Kazakhstan.

Top Uranium Players

Uranium production and marketing is largely controlled by eight companies. In 2017, these companies marketed around 86 per cent of the world's uranium mine production: KazAtomProm of Kazakhstan occupies the top slot followed by Canadian Cameco, French Orana, Uranium One (Russian), CNNC/CGN (China National Nuclear Corporation, China

General Nuclear Power group), ARMZ (Atomenergoprom Russia) in that order.

Graphic 8.4 depicts top eight companies along with their marketed uranium figures in 2013.

Most mines are found predominantly in Canada, Australia and Kazakhstan. Other countries include Niger, Russia and Namibia. Mines in these six countries alone account for over three-fifths of total global output. The table shows leading mines along with country, company, type, production and percentage share.

McArthur River Mine, Canada

The McArthur river mine is an underground uranium mine owned by Cameco. The mine is located in Saskatchewan, Canada, about 50 miles away from the Key Lake Mill in the Athabasca Basin; it contributes about 13 per cent of the world's uranium.

This is the highest-grade uranium deposit in the world, with a proven and probable mineral resource estimate of 264.5

Graphic 8.4: Top Global Uranium Players and Marketed Uranium in 2013

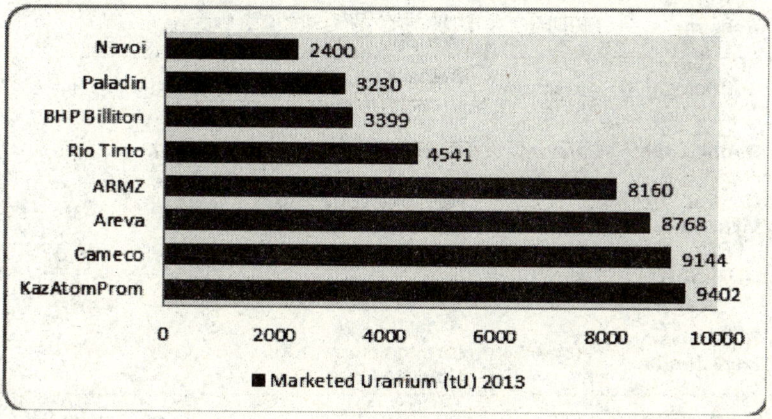

Source: WNA

World's Top Uranium Mines

Mine	Country	Main owner(s)	Type	Production (tU)	per cent of world
McArthur River	Canada	Cameco (69.8 per cent)	underground	7744	13
Olympic Dam	Australia	BHP Billiton	by-product/ underground	3399	6
SOMAIR	Niger	Areva (63.6 per cent)	open pit	2730	5
Tortkuduk (est)	Kazakhstan	Katco JV/ Areva	ISL	2563	4
Ranger	Australia	ERA (Rio Tinto 68 per cent)	open pit	2510	4
Priargunsky	Russia	ARMZ	underground	2133	4
Budenovskoye 2	Kazakhstan	Karatau JV/ Kazatomprom-Uranium One	ISL	2115	3
Langer Heinrich	Namibia	Paladin	open pit	2098	4
Inkai	Kazakhstan	Inkai JV/Cameco	ISL	2047	3
Rossing	Namibia	Rio Tinto (69 per cent)	open pit	2031	3
South Inkai	Kazakhstan	BetpakDala JV/ Uranium One	ISL	2030	3
Rabbit Lake	Canada	Cameco	underground	1587	3
Central Mynkuduk	Kazakhstan	Ken Dala JSC/ Kazatomprom	ISL	1559	3
COMINAK	Niger	Areva (34 per cent)	underground	1508	3
Budenovskoye 1and3	Kazakhstan	Akbastau JV/ Kazatomprom-Uranium One	ISL	1499	2
Top 15 total				**37,554**	**64 per cent**

*Data Source: WNA

**ISL-In Situ Leach

million pounds of uranium oxide. The McArthur river deposit was discovered in 1988 and mine construction began in 1997. Between 2000 and 2013. McArthur river operations yielded 250.6 million pounds of triuranium octoxide, an ore of uranium. In 2012, McArthur river was the world's largest uranium-producing mine, accounting for 13 per cent of world uranium production. Operations have been halted since 2017 owing to low international prices.

Olympic Dam Mine, Australia

The Olympic Dam Mine in Australia, owned by BHP Billiton, is a by-product and underground mine that produces 6 per cent of the world's uranium. Olympic Dam is primarily a large iron oxide-copper-gold deposit that produces copper, uranium, gold and silver. The mine is located at Roxby Downs, which is about 350 miles north of the city of Adelaide.

An interesting fact about the Olympic Dam Mine is that it is the largest-known single deposit of uranium in the world. The deposit was discovered in 1975 and the mine was opened in 1988 by WMC Resources. In 2005, BHP took over WMC Resources and became the owner of the mine. Processing procedures at Olympic Dam have recently improved, with uranium recovery from solution increasing to 97 per cent from 90 per cent. All uranium produced at the mine is exported.

Ranger Mine, Australia

Owned by Energy Resources of Australia, this is an open-pit mine that produces 4 per cent of the world's uranium. Located in the middle of Kakadu National Park, about 143 miles east of Darwin, the deposit was discovered in 1969. Production began in 1980. Energy Resources of Australia, a subsidiary of Rio Tinto, is also owned in part by Cameco, Cogema and a holding company representing Japanese utilities, among others. Uranium recovery at the mine ranges between 91.5 and 93 per cent. It is the second-largest uranium mine in the world, and the only one of the top three with an open-pit structure. Uranium oxide sales by Energy Resources of Australia are all exports to energy utilities in Japan, South Korea, Germany and France.

Global Recoverable Uranium Resources

Uranium resources are around 1 per cent of total global energy resources. The planet's largest reserves are buried in Australian mines totalling a huge 1,661,000 tU which comes to around 31 per cent of total world's reserves. Next comes Kazakhstan with resources of 629,000 tU. Russia, Canada and Niger occupy third, fourth and fifth place respectively. Total global recoverable resources are pegged at 5,327,200 tU. It is interesting to note that world's currently known uranium resources are largely confined (around 97 per cent) to merely 14 countries, remainder three per cent is shared by a few other countries in smaller quantities. The infographic shows resources owned by the top 14 countries.

Technological Developments and Upgradations

Interest in nuclear energy remains unabated despite sporadic hiccups, which is largely evident in continuously improved reactor designs. Newer advanced reactors now being built have simpler designs which reduce capital cost. They are more fuel efficient and are inherently safer. Generation III reactors are now in vogue and Generation IV are on the cards. Four

GI 8.5: Global Known Recoverable Uranium Resources 2011

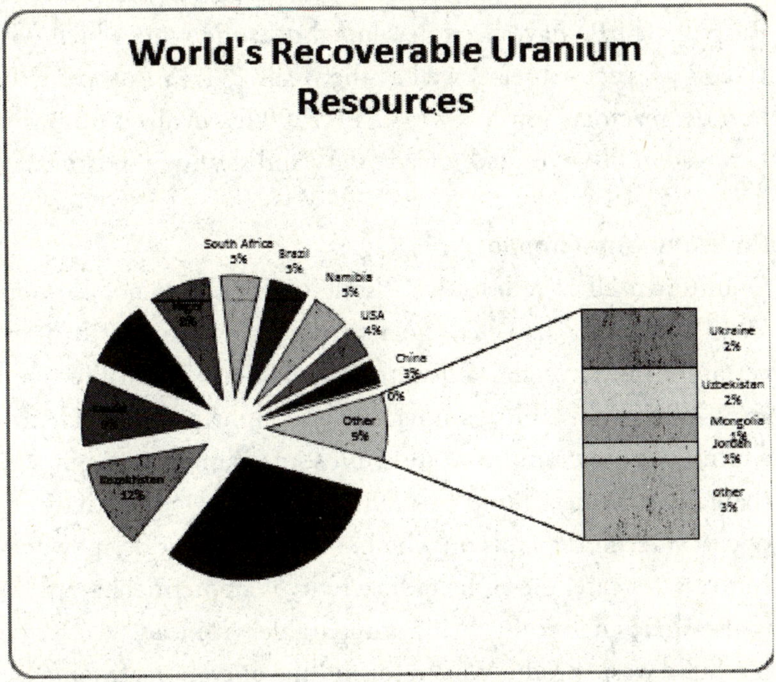

*Data Sources: NEA, IAEA andWNA

advanced reactor types were developed in the US; two of them fall into the category of large 'evolutionary' designs which build directly on the experience of operating light water reactors. These reactors are in the 1,300 mw range. One is an advanced boiling water reactor (ABWR) derived from a General Electric design and now promoted both by GE-Hitachi and Toshiba as a proven design, which is in service. The other type, System 80+, is an advanced pressurised water reactor (PWR).

India is developing the Advanced Heavy Water Reactor (AHWR) as the third stage in its plan to utilise thorium to fuel its overall nuclear power programme. China has developed ACC1000 and ACP1000 reactors. Kazakhstan has been developing a small size reactor VBER300. Fast Breeder Reactors (FBR) have been developed over the years which use U-238 as reactor fuel as well as the fissile U-235 isotope used in most reactors. South Korea's APR1400 has evolved from the US System 80+ with enhanced safety and seismic robustness.

Environmental Impact

Uranium itself is radioactive, though with the major isotope U-238 having a half-life equal to the age of the earth; it is certainly not strongly radioactive. U-235 has a half-life one-sixth of this and emits gamma rays and alpha particles. Hence a lump of pure uranium would emit some gamma rays, but less than those from a lump of granite. Its alpha radioactivity in practical terms depends on whether it is a lump or dry powder. In the latter case, the alpha radioactivity is a potential hazard. It is also toxic chemically, being comparable with lead.

Solid waste products from the milling operation are tailings. They comprise most of the original ore and they contain most of the radioactivity in it. In particular, they contain all the radium present in the original ore. When radium undergoes natural radioactive decay, one of the products is radon gas. Radon and its decay products are radioactive. Gamma radiations may pose a health hazard to those working close to high-grade ores. It comes principally from radium in the ore.

Phasing out Nuclear Power

Sweden was the first country to begin a nuclear phase-out in 1980, largely influenced by the Three Mile Island accident, followed by Italy (1987) and Belgium (1999). Austria and Spain have enacted laws to cease construction on new nuclear power stations. In the wake of the March 2011 Fukushima nuclear meltdown, Germany has permanently shut down eight of its seventeen reactors and pledged to close the rest by the end of 2022. Italy voted overwhelmingly to keep their country non-nuclear. Switzerland and Spain have also banned the construction of new reactors.

The share of renewable electricity in Germany rose from 6 per cent to nearly 25 per cent in just ten years. On sunny and windy days, solar panels and wind turbines now increasingly supply up to half the country's electricity demand, which no one expected just a few years ago. Recent estimates suggest that Germany will once again surpass its renewable electricity target and have more than 40 per cent of its power from renewables by 2020. Germans want clean energy, and a lot of them want to produce it themselves. The Renewable Energy Act guarantees priority grid access to all electricity generated from renewables and is designed to produce reasonable profits. By 2013, more than half of investments in renewables had been made by small investors. Large corporations, on the other hand, have invested relatively little so far. The switch to renewables has greatly strengthened small and midsize businesses, and it has empowered local communities and their citizens to generate their own renewable energy. Across Germany, a rural energy revolution is underway. Communities are benefiting from new jobs and increasing tax revenues, which has become even more important after the debt crisis in the euro zone. Around 380,000 Germans now work in the renewables sector.

Germany's energy transition is not only about switching from nuclear and coal to renewables in the electricity sector. Electricity only makes up roughly 20 per cent of German energy demand, with roughly 40 per cent devoted to heat and 40 per cent to transportation. Most public attention has now focused on the power sector, with the nuclear phase-out and the switch to wind power and solar power making headlines. But in fact, Germany has emerged a leader in 'passive houses' which make heating systems in homes largely redundant. Unfortunately, however, renovation rates are too low for the tremendous efficiency gains from passive house construction to be fully effective. In addition, Germany has not expanded its district heating networks, which allow waste heat from power generators to be used productively, as fast as its neighbours in Austria and Denmark.

Perhaps the greatest challenges lie in the transportation sector, where a number of options are being looked into worldwide—from electric mobility to hybrid vehicles. Germany is not a leader in such technologies. But the greatest efficiency gains will come about when we switch from individual mobility to public transport—and from large cars to small vehicles, such as electric bicycles.

Germany has now created the world's largest domestic solar PV market. German commitment and Chinese mass scale production has helped to drive down the cost of renewables worldwide. In Germany, installed system prices for solar PV plummeted by 66 per cent from 2006 to mid-2012. This shows Germany is heading towards being a nuclear-free nation even before the tryst year of 2022.

Switzerland's federal government has already decided to phase out the five nuclear reactors which generate 40 per cent of the country's electricity by not replacing them with new nuclear capacity at the end of their anticipated working lives.

The decision came in response to the accident at Fukushima. Swiss nuclear power could be phased out by 2029.

A public opinion poll conducted by *DemoScope* reflects that Switzerland's policy is not fully supported by the Swiss people. Some 73 per cent want Switzerland to be able to produce all of its electricity. Some 88 per cent of respondents said that the country's energy transition must not endanger its security of supply.

Renewed Global Interest in Uranium

Global interest is once again on the rise in this silvery-white metallic chemical element. It appears the world is gradually forgetting scourges inflicted by Fukushima. Recent reports indicate Japan is close to restart all of its closed NPPs due to mounting pressure of nuclear/industrial syndicates. Majority of nations are ramping up their nuclear operations.

As China is dependent on uranium imports to meet its growing nuclear power demand, it is now scouting for new nuclear resources in far-off lands as well as back home. China National Nuclear Corporation (CNNC) is poised to speed up overseas uranium mining exploration. The country's leading nuclear energy developer is to focus on Australia, Africa and Central Asia to meet its growing demand for raw material. The company runs more than 40 per cent of China's nuclear sites.

Recently, a large uranium deposit was discovered in North China. This discovery was significant to boost domestic uranium supplies and ensure energy sources for developing nuclear power in the country.

In 2008, China discovered its first 10,000-tonne leaching sandstone type uranium deposit in the Yili basin in the northwestern Xinjiang Uygur autonomous region.

China is a major importer of uranium. The country has around 15 operational nuclear reactors, with plans to expand its nuclear industry. Another 27 reactors are under construction near coastal areas.

China also lifted its freeze on new nuclear projects which had been put in place after Fukushima. The World Nuclear Association has reported that China will be using 20,000 tonnes of uranium a year by 2020, about a third of the global output in 2009. The country will need to source the rest from the global market, which could set the market soaring, in order to shift away from coal as a fuel source.

India Craving for Uranium from Abroad

India is not a signatory to both Nuclear Non-proliferation Treaty (NPT) and the Comprehensive Test Ban Treaty (CTBT). This poses a substantial hurdle in smooth supply of nuclear fuel and technology transfer. India has however, been exploring all energy sources from overseas as local supply is far short of its growing economy's gigantic needs. Uranium mining in Andhra Pradesh was halted for years by environmental activists but was finally restored. The state's uranium ore is nearly five times richer than in India's old mines at Jaduguda. Though Jaduguda ore contains merely 0.06 per cent uranium, and Andhra Pradesh's ore renders 0.3 per cent. Internationally, commercial ore yields go far higher up to 15 per cent. Hence Indian ores are quite uneconomical.

India as of now has civilian nuclear agreements with France, Mongolia, Argentina, Canada, South Korea, Russia, Kazakhstan, the US and Namibia. New Delhi has signed an agreement in January 2019 to procure uranium from Uzbekistan on long term basis.

Canada had stopped civil nuclear cooperation with India after New Delhi detonated a nuclear device in 1974, which Canada believed was developed using a Canadian nuclear reactor. Decades later, the India-Canada Civil Nuclear Cooperation Agreement was signed in 2010 but an 'Appropriate Arrangement Agreement' to enable Canada sell its uranium to India could be signed only in April 2013. India is expected to purchase $650 million worth of Canadian uranium annually. Ottawa plans to treble its uranium exports by exporting uranium to India and China to increase its earnings and jobs. Currently, most of the demand is met either from Russia or Kazakhstan while the remainder comes from domestic sources.

Australia holds the planet's largest known recoverable uranium reserves—about 31 per cent—which energy-starved India wants to access so it can expand its nuclear power industry. Canberra in recent years reversed course on its longstanding position against selling uranium to nations that have not signed the NPT. India developed nuclear weapons outside the international accord.

Eager to support the uranium-mining industry and deepen strategic ties with Asia's largest democracy India, the Liberal-National government in Canberra worked to speed up the pace of uranium trade talks with New Delhi. Australian Prime Minister Tony Abbott's administration had signalled it could be open to easing some of Canberra's past demands for assurances that Australian-origin uranium not be diverted to nuclear arms efforts. In September 2017, India received its first shipment of uranium after signing a deal in 2014.

THORIUM: PROMISING ALTERNATIVE

After decades of relative obscurity, thorium has been eliciting increasing interest afresh in recent times which is widely distributed in the Earth's crust. Thorium, a slightly radioactive metal that occurs in rocks and soils, may hold significant promise as an alternative to uranium in the nuclear energy sector. Thorium is usually found in small quantities in sand, soil and rocks everywhere. Large reserves exist in China, Australia, the US, Turkey, India and Norway. A US Geological Survey found thorium in Montana, Idaho, Colorado, the Carolinas, Florida and Georgia.

As global energy consumption increases, thorium is being looked at as a possible alternative to uranium to provide safe and abundant nuclear power at a reasonable cost. A breakthrough was made in 2013, when a private Norwegian company, Thor Energy, began to produce power at its Halden test reactor in Norway using thorium. Nuclear giant Westinghouse, a unit of Toshiba, is part of an international consortium that Thor Energy established to fund and manage the experiments. As part of its ongoing research into thorium as a nuclear fuel, Thor Energy created an international consortium that is charged with funding and managing thorium experiments.

Firms in the US, Australia and the Czech Republic are also working on thorium reactor designs and other elements of fuel technology using the metal. In the US, Flibe Energy is developing a thorium reactor based on designs developed in the 1960s by the US Department of Energy's Oak Ridge National Laboratory in Tennessee. China is also developing different types of thorium reactors.

Currently, China is constructing two thorium-based

reactors in Gobi desert and completion is expected this year only (2020). China expects to put thorium reactors in commercial use by 2030.

A completely different approach being explored in China and America is the liquid fluoride thorium reactor (LFTR). This technology was developed to prototype stage by the American government in the 1960s but was then shelved in favour of uranium. Proponents of this technology argue that its potential is far greater than the water-cooled approach, partly because LFTRs could extract a higher proportion of energy from thorium fuel.

Unlike uranium, thorium can't be split to make a nuclear chain reaction, in other words, it isn't fissile. However, if it is bombarded by neutrons from a fuel that is fissile, like U-235 or plutonium-239, it converts to U-233, itself an excellent nuclear fuel. After the process begins, it's self-sustaining—fission of U-233 turns more thorium nearby into the same nuclear fuel.

Thorium might be an appealing alternative to uranium to many countries. It is cheaper and rather abundant. During a thorium-powered nuclear reaction, most of the thorium itself is consumed, which leads to less waste, most of which is rendered non-hazardous in 30 years. The most dangerous nuclear waste material currently in use must be stored for 10,000 years, by way of contrast. Furthermore, 1 metric tonne of thorium is equal to 250 metric tonnes of uranium in terms of efficiency in a water reactor.

Extraction of thorium would be cheaper per unit of energy than extraction of uranium because it is present in higher concentrations by weight. Moreover, it is utterly impracticable to put this into weaponry as it contains no fissile isotope. As thorium is not fissile on its own, reactions could be stopped in case of emergency. Thorium can also be used together with conventional uranium-based nuclear power generation.

NUCLEAR POLEMICS: VIEWS & COUNTERVIEWS

Nuclear debate assumes many hues depending upon geography. In France and Ukraine, where lion's share of energy needs are met by nuclear power, it may skew towards more use whereas in Japan which recently experienced the Fukushima meltdown, the debate would tend to tilt greatly against it. However, around the world, its proponents maintain that nuclear power is an economical, safe and clean form of energy. On the other hand, opponents often cite industry disasters, problems of nuclear waste, occupational hazards and links to chronic and sometimes fatal diseases. Massive anti-nuclear protests occurred throughout the 1970s and '80s that spilled over on a smaller scale into the ongoing century. It is impossible for an issue of this magnitude to be resolved one way or the other.

Many arguments could be made in favour of or against nuclear power. Here's a list of views and counter-views.

Proponents' Views

✓ **Non-smoking energy:** Nuclear Power Plants have a distinct edge over coal/fossil fuel based plants. Unlike fossil fuel plants, which spew massive emissions of carbon dioxide into the atmosphere, NPPs don't produce emissions at all. Electricity is created by nuclear fission (splitting atoms). The iconic images of white plumes rising from cooling towers show nothing more than steam. Nuclear power is considered carbon-free and capable of producing more electricity than other renewable like solar and wind.

- ✓ **Economical and cheap:** Proponents of nuclear power often measure its economic prowess in kilowatts produced by unit combustion of various fuels and the spent amount. NPPs produce more kilowatts than coal, wind or solar for fewer bucks. As more plants are built, it's expected that construction costs will come down, making the price of nuclear-generated electricity that much more attractive. With construction come numerous jobs which make a strong case in the current economic climate.
- ✓ **Safe and sound technology:** The technology is getting safer over time. Fukushima was an old plant and meltdown is much less likely in the latest generation of nuclear reactor designs. In addition, earthquakes and tsunamis of the sort that caused the Fukushima disaster are much less common elsewhere.
- ✓ **Viable alternative to conventional fuels:** Uranium is a viable sustainable fuel source and the world has enough of uranium lasting for another 100 years. The alternatives to nuclear are coal and natural gas—including unconventional gas resources—and these would be in the long run much more polluting and damaging than nuclear.
- ✓ **Immense Potential:** Nuclear fission might potentially solve all of our energy needs. It's a valuable area of research that could guarantee abundant clean energy, so it's worth investing in the technology, continuing to use it and not abandoning it.
- ✓ **Less land requirements:** NPPs require the least acreages per unit. To generate the equivalent of a 1,000-mw plant, a single wind farm would require approximately 150,000-180,000 acres of land; a solar photovoltaic park would use 54,000 acres. By comparison, a typical NPP utilises merely 200-400 acres.

OPPONENTS' COUNTERVIEWS

- **Unsafe and highly risky:** It's well understood that nuclear accidents can be both highly risky and extremely costly. The meltdown at the Fukushima NPP was the worst since Chernobyl in Ukraine 25 years earlier, and caused an estimated $74 billion of financial losses in Japan alone.
- **Renewables a better choice:** Renewables are ready to take over from nuclear. In fact, we could be producing 100 per cent of our energy from renewables by 2050, and the technology is already ready for the market—particularly if the subsidies for fossil fuels and nuclear are cut. Furthermore, if we don't start using renewable now then we may never make the switch, so this is the chance to take that first step.
- **Radioactive waste issue:** There is the significant issue of radioactive waste, which isn't biodegradable and is extremely dangerous. Most plants store nuclear waste in steel-lined concrete basins filled with water, where it remains radioactive for thousands of years.
- **Not that cheap:** Critics argue that the cost benefits aren't so clear-cut. While the electricity seems cheaper up front, the exorbitant costs of building and maintaining plants must be added into the equation—something industry experts rarely do. Long-term storage of nuclear waste is expensive and dangerous. Next, add the expense of finding and retaining skilled labour. When all is said and done, the claim that nuclear power is more cost effective than alternatives like wind, solar or even coal becomes a little cloudier. Public investment in nuclear energy far outstrips investment in renewables. Nuclear plants are also expensive to set up and decommission, and the costs of storing radioactive waste (effectively indefinitely) also have to be considered. If the

money pumped into nuclear had been spent on renewables, then the pay-off would have been much greater per unit invested.

CURRENT TRENDS

A July 2017 report by the IAEA on 'International Status and Prospects of Nuclear Power' said that some 28 states without NPPs are considering, planning or starting them. A further 20 countries have expressed an interest in nuclear power.

Belarus, alongside the UAE, is set to complete construction of its first NPP this year in 2020. They will soon be joined by reactors currently under construction in Turkey and Bangladesh, with plans for new construction well advanced in countries such as Egypt and Uzbekistan.

On a global scale, 15 countries—including China, India and Russia—will conclude the development of nuclear power projects, which are currently under construction, in the next few years beyond 2020.

According to the World Nuclear Association, there are 440 nuclear power reactors in the world, amounting to a combined capacity of 400 gwe, while more than 100 have been ordered. Estimates say that in 2018 these produced 2,563 twh of electricity, roughly 10 per cent of the world's electricity.

Between 2020 and 2026, 48 power reactors will be built globally. The peak will take place between 2020 and 2022, with 32 reactors due to begin operations. China will lead the way with 12 plans already underway, followed by India and Korea.

Belarus's first NPP Ostrovets1, situated 130 km north-west of the capital Minsk, is set to be operational in 2020. Funded by Russia, Ostrovets1 will consist of a VVER-1,200 pressurised reactor, producing 1,200 mwe of energy. A second reactor, Ostrovets2, will have the same capacity and will be built by 2021.

Just like Ostrovets1 for Belarus, Barakah1 will be the UAE's first nuclear plant, as well as the first one in the Arab world. The project, a joint venture between Saudi Nawah Energy and South Korean utility company Kepco will also become operational in the year 2020. Barakah1 will generate around 11 twh and, alongside three other reactors, will produce 134.6 twh of electricity meeting 25 per cent of the country's needs.

On the other hand, Germany, Korea, Belgium and Switzerland are planning to phase out nuclear power. France, the nuclear superpower, is planning to significantly reduce its reliance on the technology. In the US, Japan, the United Kingdom, Canada and Sweden, because of economic and technological issues, nuclear power is losing ground to some extent yet there is no voluntary policy in place to reduce nuclear power. In Russia, no leader in the fight against Climate Change, the fate of nuclear power does not seem so gloomy–without being much brighter either. Similarly China has augmented its capacity from 191 twh capacity to 277 twh capacity in just seven years (2011-2017).

Current trends indicate the countries which at present do not have nuclear energy in their energy basket would continue to embark upon installing new reactors with technological knowhow from the advanced countries.

Cautious Approach and Smooth Transition

The world is deeply divided on nuclear energy—there are fierce opponents and at the same time there is no dearth of nuclear lovers either. But sooner or later, the world has to renounce the nuclear obsession, having suffered so much due to nuclear disasters and through nuclear weapons as witnessed in Hiroshima and Nagasaki. NPT and CTBT are both discriminatory in their present form—they need suitable modifications to usher in a nuclear-free world. South Africa has destroyed its entire nuclear arsenal. Why can't other countries follow suit? It is true an immediate cessation of nuclear power production is not feasible and practicable but a smooth transition is surely possible. Germany is on the trajectory of marvellous transition through its pragmatic Energiewende programme and likely to phase out all its NPPs. What is required is strong political will as exhibited by the German government. Old plants could be retired after running their useful economical life and no fresh approvals should be granted. Retired plants could be put into reuse by some modifications to utilise biomass or bio-waste fuels. Running plants should be encouraged to use thorium instead of uranium as that would not be so damaging in case of a tsunami. Thorium-based reactors are relatively safer as the reaction can easily be stopped and operations need not take place under extreme pressure. Thorium reactors produce far less waste and the generated waste is much less radioactive and short lived as compared to uranium.

Sand Mining Mania

People have been removing natural resources for a very long time but this will continue to increase. We are lagging in our knowledge of the consequences of sedimentary exchanges between the shore and the foreshore. Sandbars are moving, but we are not able to quantify these fluxes. As a result, storms can be devastating for communities living on the coast. Men shouldn't try to play sorcery's apprentice.

—Eric Chaumillon, Marine Geologist

HEIGHTENED INFRASTRUCTURE development in developing countries and the construction industry's demands in other parts of the globe have resulted in dramatic spurt and excessive mining of this scarce, vital and critical natural resource. Illegal sand mining is more conspicuous in Asia and Africa. Sand is the second most important natural resource in the world after water, used in a number of products, amenities and utilities from glass to microprocessors, vacation homes to sky-rocketing buildings, railway tracks to highway roads. Since it constitutes an inevitable and indispensable ingredient in the composition of concrete, it is the object of such high craving and lofty greed!

Beaches are now severely affected. More than 70 per cent of the world's beaches are now on the verge of vanishing. Beaches constitute the first line of defence that coastal communities have against storm impacts, which are becoming more severe and more costly in terms of both dollars and lives as a result of increasing beach development. Almost 60 per cent of the world's population lives on 10 per cent of the planet's land—on the coast. Sand is being mined from beaches for use in concrete to support the booming construction industry. Smuggling of sand from one country to another is predominantly prevalent in South East Asia.

Illegal sand mining has been documented in at least five dozen countries spread over five continents. The curse is palpable in Morocco, India, South Africa, Benin, Liberia, Grenada, Jamaica, Malaysia and Indonesia. Beach sand mining has also been documented in Mexico, Canada, and Australia and elsewhere.

For many years now, the Caribbean has been a victim of the illegal sand mining and consequently beaches from Jamaica to Barbuda have drastically shrunk. The pillaging of sand is a growing practice the world over. Taken by hand, three or four metres deep in the Maldives archipelago and transported on a donkey, or sucked up by huge sand boats in Asia, coastal sand mining—both authorised and unlawful—is exploding.

Sand quarries are already over-exploited, or subject to stringent environmental regulations in developed countries, they are also becoming more expensive. Sand mining has widely brutal repercussions, not only for the deep-sea ecosystem and fishing, but also for the coast, where erosion has been accelerating. The receding of the coastline is proving catastrophic, which is amply attested by increased storms, floods and the infiltration of seawater in freshwater bodies. Some governments have started fighting against this phenomenon. Regulations have been drafted and are now enforced in industrialised economies but that is not yet the case in the rapidly growing economies of South-East Asia and many other countries.

A big part of the African continent is affected by sand theft. Benin has closed its coastal deposits. But in Sierra Leone, which is under the impetus of reconstruction after years of civil war, trucks have unlawfully invaded the beaches. Senegal has also banned extraction of sand from most of its coastline. But these measures are rarely respected. Around Dakar and Saint-Louis, coastal protection committees struggle to stop the pillage. Local organisations are multiplying, but have not been able to unite their efforts. People living on the periphery of the system are forced in many poor countries to collect sand from dawn until sundown.

Sand mining is even more conspicuously present in Asia, where demographic pressure and economic development force cities to extend toward the sea. The island of Singapore has grown by 20 per cent since the 1960s by taking sand from neighbouring countries. In response, Indonesia, Malaysia, Cambodia and Vietnam have banned the export of sand. Smugglers have taken over the market in Cambodia. The artificial islands of Dubai required a massive quantity of sand, favoured by the low cost of shipping. Desert sand, too thin and too round, was not suitable. The Palm and the World islands of Dubai required at least 600 mt of sand, which is more than France's annual sand production. The legality of sand mining varies from country to country.

Genesis of Abnormal Spurt in Demand

Though little public data exists about extraction volumes, sources and uses we know that, with some exceptions, most sand and gravels extracted from natural environments are consumed regionally because of the high costs of transport.

Aggregates (crushed rock, sand and gravel) account for the largest volume of solid material resources extracted from systems globally (Peduzzi, 2014; Beiser, 2018). An estimated 40-50 billion mt is extracted from quarries, pits, rivers, coastlines and the marine environment each year (Peduzzi, 2014; Beiser, 2018). The construction industry consumes over half of this volume annually (25.9 billion tonnes to 29.6 bt in 2012: UNEP, 2014) and could consume even more in future (Beiser, 2018).

For example, two-thirds of global cement production occurs in China (58.5 per cent) and India (6.6 per cent) (United States Geological Survey, 2017; Gavriletea, 2017). These two countries also lead in global infrastructure construction (Bagnasco *et al*, 2015; Global Construction 2030; World Economic Forum 2019) as these nations look to rapidly transition their populations out of poverty.

The Global Aggregates Information Network (GAIN) suggests that at one end of the scale, least developed regions use 3-4 tonnes/capita per year and developed regions with slow growth, flat terrain use 4-8 tonnes/capita per year (based on 2016 data). Growing economies with rugged terrain and harsh climates can use 8-16 tonnes/capita per year. China stands out for aggregates production at 14.3 tonnes per capita per year. In total, China, India and Asia represent 67 per cent of global aggregates production.

The industry outlook to 2030 is one of strong growth in Asia, particularly India, Africa and Latin America (Gavriletea, 2017). GAIN estimates global aggregates demand will likely to rise to 60 bt per annum by 2030 driven by increasing population, urbanisation and economic growth. Desert sand, though plentiful, is unusable for most purposes because its wind-smoothed grains render it non-adherent for the purposes of industrial concrete.

The trend of sand extraction can be estimated using cement production as a proxy. Five countries—China (58 per cent), India (6.75 per cent), the United States (2 per cent), Brazil and Turkey—now produce nearly 70 per cent of the world's cement. Cement demand from China increased exponentially by 437.5 per cent in the past 20 years, while use in the rest of the world increased by 59.8 per cent only. Each citizen of China is currently using 6.6 times more cement than a US citizen. Demand continues to increase with new infrastructure and renewal, renovation and rehabilitation of existing infrastructure like roads, bridges, dams and houses.

Lessons from Europe

Detrimental consequences of river sand mining were first sensed in Europe where the use of sand had started quite early. It gradually dawned that sand extraction has an adverse impact on river dynamics and ecology. European states perceived the harmful impacts of sand mining way back in the 1950s, chiefly in Italy where large-scale mining provided the raw material for the vast development of the motorway network. In France, the impact of gravel and sand mining was seen as pernicious in the late 1970s, found unsustainable in the 1980s and banned in the early 1990s.

Sand Mining near Beaches

The world's beaches are being mined for sand for a variety of uses such as aggregate in concrete, foundation and plinth fill, beach re-nourishment. This is a global phenomenon from Morocco to Michigan, Caribbean Islands to Far East Raglan in New Zealand. It is rampant along coastal Kerala and Kendrapara, Odisha in India, Carriacou (Grenada) Contonou (Benin), Raglan (New Zealand), Melkbosstrand (South Africa), Stradbroke (Australia), Tangier (Morocco), Trelawny (Jamaica), Michigan (USA) and Buchanan (Liberia).

The practice is often very destructive and poorly managed or unmanaged. The pillage of beach and dune sand is a direct cause of erosion along many shorelines. It is very damaging to the beach fauna and flora, ruinous to beach aesthetics and frequently causes environmental damage to other coastal ecosystems associated with the beach such as wetlands.

Another major impact of beach sand mining is the loss of protection from storm surges associated with tropical cyclones and tsunamis. Some communities affected by the 2004 tsunami in the Indian Ocean had higher storm surges probably due to beach sand mining resulting in fatalities. Sometimes it is difficult to tell that a beach has been mined. Sand extraction becomes difficult to recognise as the beach readjusts to a new profile after a few storms. But historic accounts of beaches in the Caribbean reveal that beaches have been narrowed considerably. Mining is particularly senseless in a time of rising sea level when sand is sorely needed as a buffer.

Global Implications of Beach Sand Mining

Millions of cubic metres of sand has been removed across

five continents. Problems extend from Asia, where critical infrastructure has been disrupted by sand mining, to the Caribbean where policy reform has swiftly followed a highly publicised theft of sand. Extensive sand mining in Morocco has resulted in tens of kilometres of stripped beaches and the mining continues southward, reducing hope of a thriving tourism-based economy.

Prominent problems caused by beach sand mining activity include:

- Destruction of natural beaches and the surrounding ecosystems they support such as dunes, wetlands
- Habitat loss for globally important and vulnerable species like turtles, shore birds, snails
- Degradation, partial or complete destruction of near shore aquatic ecosystems,
- Augmented shoreline erosion rates
- Reduced protection from storms, tsunamis and other extreme

Cartographic Depiction of Beach Sand Mining Locations around the globe

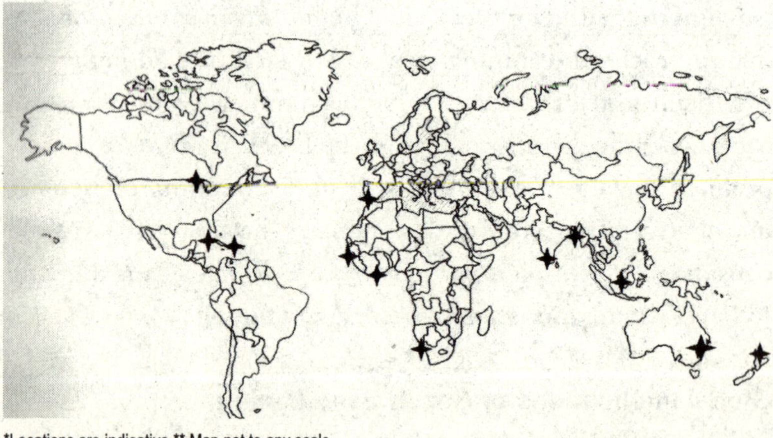

*Locations are indicative ** Map not to any scale
Infographic by Author for reader facilitation

weather events and consequential economic losses through tourist abandonment/reduction
- Loss of coastal aesthetics.

The threats posed by sand mining are made even worse by the looming prospects of a significant rise in global sea level (GSL) over coming decades. Most governments recognise the local impacts of sand mining and also know about illegal mining. However, enforcement has been problematic, tardy. Strong action often remains elusive owing to nexus of sand mafia, politicians and patrons. There has been little or insignificant pressure to check the malpractice from local and international environmental groups.

In many cases, addressing the issue of sand mining entails addressing the local issues that allow it to persist. This includes poverty, corruption, and unregulated development. In areas where beach sand mining significantly supports the local economy, care needs to be taken that local workers are given alternative means of income and builders are provided affordable substitutes for sand.

Consequences of Excessive Sand Mining

Sand is also extracted from rivers, streams and rivulets wherever these are in close vicinity of construction activity. Stream mining lowers the stream bed that may lead to bank erosion. Excessive river sand mining is a threat to bridges, river banks and nearby water structures. Sand mining also affects the adjoining groundwater system.

Depletion of sand in the estuarine bed causes the deepening of rivers and estuaries, enlargement of river mouths and coastal inlets. It may also lead to ingress of saline water from the nearby sea. The effect of mining is compounded by the effect of sea level rise. Any volume of sand exported from streambeds and coastal areas is a loss to the system.

Stream sand mining results in the destruction of aquatic and riparian habitat through large changes in the channel's geomorphology. Impacts include bed lowering, degradation, bed coarsening, depleted water tables near the stream bed and channel instability. These physical impacts cause degradation of riparian and aquatic biota and may lead to the undermining of bridges and other structures. Continued extraction may also cause the entire stream bed to degrade to the depth of excavation. It practically alters the entire river regime.

Sand mining generates extra vehicle traffic, which negatively impairs the environment. Where access roads cross riparian areas, the local environment is greatly affected.

Riparian Habitat Degradation

Stream mining may similarly have other pernicious effects beyond the immediate mine sites. Many hectares of fertile streamside land are lost annually, apart from valuable timber resources and wildlife habitats in the riparian areas. Degraded stream habitats result in diminishing of fisheries productivity, biodiversity loss and reduction in recreational potential. Severely degraded channels may degrade land and aesthetic value thereof.

Threats to Aquatic Life

All species require specific habitat conditions to ensure long-term survival. Native species in streams are uniquely adapted to the habitat conditions that existed before humans began large-scale alterations. These have caused major habitat disruptions that favoured some species over others and caused overall declines in biological diversity and productivity. In most streams and rivers, habitat quality is strongly linked to the stability of channel bed and banks. Unstable stream channels are inhospitable to most aquatic species.

Factors that augment or lower sediment supplies often destabilise beds and banks and result in dramatic channel readjustments. For example, human activities that accelerate stream bank erosion, such as riparian forest clearing or river mining cause stream banks to become net sources of sediment that often have severe consequences for aquatic species. Human activities that artificially lower stream bed level give rise to bed instabilities and subsequently result in a net release of sediment in the local vicinity. Unstable sediments simplify and therefore degrade stream habitats for many aquatic species. Few species benefit from these effects.

The most detrimental effects of stream sand mining on aquatic habitats are bed degradation and sedimentation, which can have substantial negative effects on aquatic life. The stability of sand-bed and gravel-bed streams depends on a delicate balance between stream-flow, sediment supplied from the watershed and channel form. Mining-induced changes in sediment supply and channel form disrupt channel and habitat development processes. Furthermore, movement of unstable substrates results in downstream sedimentation of habitats. The affected radius depends on the intensity of mining, particles sizes, stream flows and channel morphology.

The partial or complete removal of vegetation and

destruction of the soil profile destroys habitat above and below the ground as well as within the aquatic ecosystem, results in the reduction in faunal populations. River widening results in shallowing of the stream bed, producing braided, twisting or interweaving flow, simultaneously augmenting sub-surface inter-gravel flow in riffle (stream rapid) areas, hindering smooth movement of fishes between pools. Channel reaches become more uniformly shallow as deep pools laden with gravel and other sediments, reducing habitat complexity, riffle-pool structure, and numbers of large predatory fishes.

Water Quality Affected

River sand mining activities have an impact upon river water quality. Impacts include increased short-term turbidity at the mining site due to suspension of sediment, sedimentation due to stockpiling and dumping of excess mining materials and organic particulate matter. Enhanced erosion process both at riverbed and bank increases suspended solids in the water at the excavation site and further downstream. Suspended solids may adversely affect potential water users and aquatic ecosystems. The impact is more pronounced in case water users downstream of the site are utilising water for domestic use. Moreover, suspended solids can significantly increase water treatment costs.

Damage to Groundwater

Stream sand mining transforms the entire stretch of a river and the riverbeds are converted into large, deep pits. Consequently, the groundwater table drops considerably, leaving the nearby drinking water wells dry. Bed degradation emanating from stream mining lowers the level of stream flow and the floodplain water table in turn can eliminate water table-dependent vegetation in riparian areas, and greatly decrease wetted periods in riparian wetlands.

Recapitulation

Here's a summary of all the ill effects of excessive river sand mining enumerated above:

- Erosion of channel bed and banks, increase in channel's longitudinal slope and change in entire river morphology
- Undercutting and collapse of river banks
- Upstream erosion owing to an increase in channel slope and changes in flow velocity
- Enhanced downstream erosion
- Damage to nearby land and adjacent structures
- Affected water quality and groundwater table
- Destruction of stream habitat, disturbances of attached species and reduced feeding opportunities.

Rampant Sand Mining in Asia

In many an Asian country, illegal sand mining is often resorted to by highly unscrupulous people, gangs of criminals and close cronies of those in power. This lucrative business fetches huge booty which is later shared among these depraved extractors, corrupt officers and political lords. The whole game has become so vicious that the said nexus is like an impregnable reinforced concrete wall. Lack of clarity of rules, weak institutional mechanism to enforce extant laws, tardy and cumbersome process to bring the culprits to book, lack of awareness/sensitivity of the issue, absence of powerful united fronts opposing sand mining—all these put together make it extremely difficult to thwart and solidly arrest ongoing illegal mining in Asia and elsewhere. Let's examine some interesting countries of Asia and the extent of damage.

Cambodia

Cambodia is a country rich in natural resources yet around 36 per cent of its population still lives in poverty and over 1.5 million Cambodians struggle daily to ensure their sustenance.

Global Witness, a UK-based NGO in its 2010 report, estimated that around 796,000 tonnes of sand were being exported every month from Cambodia to Singapore. The latter, a city state, was the world's largest importer of sand in 2008. Singapore has considerably increased its landmass since 1960 mostly through land reclamation from adjacent seas. Singapore is heavily dependent upon nearby regions for meeting its sand demand spurred by construction boom and constant addition/enhancement of its skylines.

Another reliable study by the Mekong River Commission (MRC) published in 2012 indicated Cambodia extracted a whopping 18,740,000 cum of sand out of river Mekong in the year 2011 alone.

Pristine rivers of Koh Kong province, Sre Ambil, Ta Tai and Koh Por, have been invaded by large sand dredging boats and barges and humongous quantity of sand has been extracted from these rivers. The dredging along the rivers of Koh Kong has decimated the rich marine life to a considerable extent and destroyed the livelihood of the poor fishermen and villagers who largely relied on the river for eking out their livelihoods.

Laos (Lao People's Democratic Republic)

Laos is a beautiful country in South-East Asia, perfectly landlocked and dotted by mountains from east to west. Laos is rich in metal and mineral resource and therefore many global players from the rich countries of Asia and Europe are quite active here. Mighty river Mekong flows through this country and sand is extracted from this river at multiple locations. At some locations, mechanised extraction deploying big dredging machines is also resorted to. One reliable study revealed it extracted huge quantities of sand from Mekong basin area; over 4,000 cum in the area covering Vientiane–Savannakhet in 2011. Upstream of Vientiane, it extracted 87,000 cum, and 341 cum of sand was removed from the area encompassing Savannakhet-Champasak.

Laos is an opaque country where big businesses can swallow resources with minimal scrutiny. Sand mining continues to pick up speed. River Mekong has changed considerably here due to dredging operation and river banks are collapsing. Uptick in sand mining is harming delicate ecosystem of the Mekong. Many Chinese have been constructing huge buildings in Vientiane which requires mammoth quantity of sand and aggregates. Sand is also being extracted heavily from the River Nam Khan (tributary of Mekong) near Luang Prabang, Laos where new resorts are being built. China is now the largest source of foreign investment in Laos.

China

Demographic compulsions render China a voracious resources seeker from gasoline to coal, hydropower to wind power, rare earths, metals and practically every conceivable resource. Hence

China can't lag behind in extracting this critical resource of sand which is also inevitable for its skyrocketing construction industry.

The Yangzi and Pearl Rivers, provide a classic example of the severe impacts affecting the ecology, groundwater level and bank erosion. Extraction later shifted from the Yangtze riverbed to Poyang Lake after legislation protecting the river was enforced in 2000. The rate of sand extraction from Poyang Lake was estimated at 236 million cum per year in 2005 and 2006 tantamount to 9 per cent of Chinese gross demand. The extraction rate from Poyang Lake was thought to be the highest in the world at that time. As a consequence, the hydrologic buffering of the lake has decreased and biodiversity hot spots are threatened. This is extremely important because Poyang Lake is the major expansion area for Yangzi floods.

An estimated 870 million cum of sand was excavated between 1986 and 2003 in the rivers of the Pearl basin, resulting in severe lowering of river channels, bank instability and threats to infrastructure, changes in the divided flow ratio between various water courses, modification of saltwater intrusion levels and a limitation of the freshwater supply. The extraction rate was much higher than the replenishment rate, a situation aggravated because sediment fluxes depleted dramatically due to the combined effects of declining rainfall, dam construction, water diversion, reforestation and afforestation.

Vietnam

Excessive and illegal sand mining has caused severe erosion along the two main tributaries of the Serepok river in the Tay Nguyen (Central Highlands), affecting the lives of many people. In Khoi's Cu Kty commune alone the river has swallowed up around 80 ha of farmland since 2005. Several sections of banks in the province's Cu Kuin district have also been badly eroded. Provincial Natural Resource Management Department authorities say it is difficult to eradicate illegal sand mining in the Krong Ana River. The other river, the Krong No in Dak Nong Province, has become wider and deeper because of illegal sand mining. The section running through Krong No district's Buon Choa commune used to be just 20 metres wide a decade ago, but has now widened to 100 metres. More than 30 sand-mining boats operate in this section of the river, taking a total of 1,800 cum of sand every day.

Maldives

By the turn of this century, average GSL rise is expected to reach 0.26 to 0.55 metres under the best-case scenario (of 70 per cent reduction of greenhouse gas emissions), and nearly one metre under unabated increase in greenhouse gas emissions (IPCC 2013).

This problem is more exacerbating for small islands states like the Maldives where retreat options are limited. In the Maldives, a few of the largest and highest islands, such as the capital city Male are being consolidated to help ensure accommodation of the displaced population from low-lying islands. To strengthen the city, a large amount of sand is being imported, to be used in building higher towers and coastal protection in Male. The sand is taken from offshore sand islands. Ironically, the sands extracted for the protection measures in Male are leading to the lowering of some islands, increasing the need to relocate their populations.

India

India's construction industry accounted for two–fifths of the development investment in India in the past half century. India comes third after China and the US in terms of construction business, which roughly accounts for 9 per cent of its \$2 trillion economy. It has highly ambitious plans to invest \$500 billion for infrastructure growth. India consumes around 450 mt of concrete annually and its construction industry has been growing by approximately 15 per cent annually. These factors have triggered desperate legal/illegal mining in India and a sharp spurt has been witnessed both in demand and prices.

Though the sector is highly unorganised, enhanced mechanisation and advent of suction boats facilitating rapid extraction have aggravated the situation. In June 2011, Swami Nigamananda sacrificed his life after observing a four-month fast to protest sand mining along the banks of Ganga. His death served a serious warning against the reckless and proliferating sand mining in the country.

In Maharashtra, the Bombay High Court banned the extraction of sand in 2010 largely due to its adverse environmental effects. At the behest of the court, the state government came up with a new policy on sand mining empowering local communities to have a say in sand mining in their localities. The policy also banned use of suction pumps in dredging but failed to check mechanised dredging. The High Court also directed the government of Maharashtra not to award any contract or allow any sand excavation to commence across the state without prior environmental clearance. However, around a month later, the ban was lifted and many key issues remained largely unresolved.

Similarly, in December 2010, the Madras High Court, disposing various Public Interest Litigations (PILs) filed on the issue of illegal sand mining in the state of Tamil Nadu ordered the formation of a high-powered monitoring committee to prevent illegal sand extraction and ensure proper and skilful quarrying of sand from river beds. The state government of Tamil Nadu later submitted to the court that it had formed such a committee in March 2011 for a period of two years.

In February 2012, the Supreme Court of India in its landmark decision in the case *Deepak Kumar and Others Vs State of Haryana and Others* listed a set of conditions for granting permission for sand mining in the states. Among the major recommendations of this order was for mining leases for riverbed sand mining based on identification of specific stretches. The apex court made it mandatory for the Environmental Impact Assessment (EIA) report to be carried out before granting a licence to extract. Apart from this, the Court said extraction should not be done deeper than three metres below water level.

Thereafter, in a momentous decision dated 5 August 2013, the National Green Tribunal (NGT) banned sand mining from river beds without environmental clearance from the Ministry of Environment and Forests. The NGT observed that besides violations of law, mining activity is being carried out on a large scale, causing state revenue loss which may be running into lakhs of crores of rupees. The tribunal also directed deputy commissioners, superintendents of police and mining authorities in all states to ensure compliance with its order. The ambit of the verdict covered all rivers across India. The NGT order came in response to a petition that raised the issue of large-scale illegal mining activity being carried on the banks of the Ganga, Yamuna, Chambal, Gomti and Revati rivers, among others.

Extensive Mining in Rivers and along Beaches in India

When demand far outstrips supply, illegalities and corruption find their place. The complexity of the issue can be viewed in the case of Mumbai, the commercial capital of India. The metropolis alone needs at least 15,000-18,000 trucks of sand per day, while the availability is only 600-800. This huge deficit gives birth to a host of illegal activities, further compounding the problem.

In Uttar Pradesh, the sand mafia wields full sway and wreaks much havoc. No region or river is spared. Plundered regions include Vindhyachal, Purvanchal, Bundelkhand right up to the Ganga-Yamuna plains, the rivers being Chhoti Gandak, Gurra, Ghaghara, Ganga, Yamuna, Ken, Betwa, Rapti and Gomti. The Allahabad High Court had imposed a ban on the use of machines in sand mining on the riverbeds of Ganga and Yamuna in Allahabad and Kaushambi districts.

Considerable damage has already been done to Kerala's major rivers Achankovil, Pampa and Manimala. Other hotspots in Western India include Narangi, Bhayander, Raigad district, Roha, Mahad, Kihim and Awas beaches. Significant quantity of monazite sand has been removed from Tamil Nadu and Kerala beaches.

Viable Alternatives to Sand

Growth is an inevitable proposition so we must think of replacing sand by various tested materials without compromising on the strength of the concrete. Various part of the world have already discovered such materials and are successfully utilising them. There has been a dramatic rush in research for alternatives to sand, as most governments are proclaiming ban on sand mining or restricting the same. The need of the hour is to popularise these materials, dispel unfounded fears and vapourise inherent inertia. Let's have a look at the available alternatives, their suitability and viability.

M-sand (Manufactured Sand)

M-sand is a promising and viable alternative to natural river sand as it doesn't contain either silt or organic impurities mostly found in natural sand. Some states of India have started the use of M-sand. The ambitious Pune-Mumbai Expressway was built with artificially manufactured M-sand, which is manufactured by crushing stones in large crushers followed by other statutory engineering processes. Its particles are cubical, unlike natural sand which has spherical particles. The cubical shape of the particles helps make concrete more cohesive and the perfect gradation ensures fewer voids. Hence, M-sand has slightly higher compressive strength and better water tightness than conventional sand.

Coal Bottom Ash

Coal Bottom Ash (CBM) is granular, glassy and greyish in appearance. The coal burning process results in production of about 80 per cent fly ash and 20 per cent CBA. The latter is collected from furnaces that burn coal. Though the fly ash has found adequate takers, CBA remains largely unutilised.

In India, around 100 mt of fly ash is produced annually and 15-25 mt of CBA remains unutilised posing a great environmental threat as well. Some countries lack suitable land sites for the disposal of such ash.

Thermal power plants are still being constructed in major developing countries across the globe so there would not be any dearth of this by-product. Its use as a replacement for sand up to 30 per cent in concrete has been demonstrated in lab tests.

Copper Slag

Copper slag is a known abrasive and is chiefly for surface blast cleaning the world over. In some geographies, it is also used as building material in the form of blocks where copper smelting facilities exist. At least 5 per cent of sand can be replaced by copper slag in concrete and around 30-40 per cent in pavement grade concrete. Over 30 mt of copper slag is produced annually the world over. This can be very well utilised. Moreover; supply of this slag is expected to remain steady.

Foundry Sand (Used and New)

Foundry Sand (FS) also known as metal-casting sand, is normally high quality silica sand, a by-product of the ferrous and non-ferrous metal casting industry. Sand has been used for centuries as a moulding material because of its unique engineering properties. In modern foundry practice, sand is typically recycled and reused through many production cycles. Large quantities of sand are used in metal casting industries around the globe. A significant percentage is discarded annually and is available for recycling and other uses.

Extensive research has shown that the compressive strength of the concrete clean (new) foundry sand to replace sand by 25-35 per cent has the same strength that of a mix containing 100 per cent natural sand. However, if discarded (used) foundry sand is used (25-35 per cent), then the compressive strength of the concrete gives 20-30 per cent lower values than concrete without used foundry sand.

Granulated Blast Furnace Slag

Studies conducted so far amply attest to the use of Granulated Blast Furnace Slag (GBFS) in concrete with higher sand replacement. As much as 75-80 per cent of sand in concrete can be replaced with assured strength. This is widely available from metal industries as a by-product. Indian iron and steel industry alone produces over 10 mt of GBFS and there would not be any let-up in supply of this slag with consistent steel demand being recorded in many developing countries.

Crushed Waste Glass

Crushed glass is perhaps the only material as fine aggregate which can replace sand by 100 per cent. Crushed glass powder also exhibits pozzolanic properties depending upon fineness of the powder. Concrete containing glass as fine aggregate can achieve comparable strength to that of natural sand aggregates. However, the concrete containing glass gives the visual appearance of lack of physical bond. Further research is warranted for this promising alternative.

Quarry Rock Dust

Quarry Rock Dust (QRD) is a residual waste material after the extraction and processing of rocks to form fine particles less than 4.75 mm. QRD is easily available from crushers. Crushers produce at least 20 per cent quarry rock dust as waste material. QRD finds its use largely in the highways as a surface finishing material. It is also used for manufacturing of hollow blocks and lightweight concrete prefabricated elements. Some studies have found the compressive strength, flexural strength and durability of concrete made of QRD is nearly 10 per cent more than conventional concrete.

The ideal percentage of the replacement of sand with quarry dust would be in the region of around 55-75 per cent. Even higher replacement is possible by addition of fly ash in appropriate percentage.

Construction and Demolition Waste

Construction and Demolition Waste (CDW) arises from activities such as the construction of buildings and civil infrastructure, total or partial demolition of buildings and civil infrastructure, road planning and maintenance.

In Europe, CDW accounts for approximately 25-30 per cent of all waste generated and consists of numerous materials, including concrete, bricks, gypsum, wood, glass, metal, plastic, solvents, asbestos and excavated soil. Concrete and hence sand forms a significant quantity to be retrieved out of CDW. There is a high potential for recycling and re-use of CDW, since some of its components have a high resource value. In particular, there is a re-use market for aggregates derived from CDW waste in roads, drainage and other construction projects. In addition, technology for the separation and recovery of construction and demolition waste is well established, readily accessible and in general inexpensive.

Of late, the European Commission has issued directives to all its member states that they shall take the necessary measures designed to achieve retrieval of a minimum of 70 per cent (by weight) of non-hazardous construction and demolition waste by 2020.

Some Scandinavian countries in Europe stipulate more rigorous rules. No fresh construction materials are allowed until and unless all the discarded materials and debris are fully recycled. Such laws should be emulated in other countries to reduce the tremendous pressure of legal/illegal sand mining across the world.

Tackling the Menace

A mammoth gap exists between the magnitude of the problem and public awareness of the issue. Moreover, the absence of global monitoring of aggregates extraction certainly contributes to the gap in knowledge which invariably translates into highly deficient and unfocussed actions. A recent French/English documentary *Sand Wars* (2013) by Denis Delestrac brings to the fore the global menace of sand mining. This award-winning eco-thriller raises awareness of the blazing issue and depicts the mafia, contractors, sand smugglers and unscrupulous property developers; the entire bunch grossly indulge in this new gold rush. It also portrays environmentalists and local populations struggling to reverse the threat to the future of this vital resource that we all take for granted.

As the issue of sand mining is now assuming international dimensions, there is an urgent need for in-depth research especially oriented for South-east Asia. The implementation of a monitoring mechanism regarding global aggregate extractions and trade would shed light on the magnitude of this issue and bridge the current data and knowledge gap. This would also raise this issue on the political agenda and perhaps lead to an international framework to improve extraction governance. Though UNEP came out with a report titled 'Sand & Sustainability' in 2019, an international regulatory framework to control illegal sand mining remains elusive. The report also speaks of absence of reliable extraction data as most of the extraction remains illegal and done stealthily and quite often goes unreported. The current levels of political concerns are far from compatible with the dire urgency of the situation!

Avaricious Eyes on Forests

Destroying rainforest for economic gain is like burning a Renaissance painting to cook a meal

—Edward O Wilson, American Biologist and Naturalist

The world has lost more forest area than the size of South Africa since 1990 viz.
we have lost equivalent to 1,000 football fields of forests every hour

—World Bank World Development Indicators 2016

The world lost nearly 13 mha of forest—an area roughly the size of England—every year between 2000 and 2010

—World Resources Institute

ANCIENT INDIC and Buddhist scriptures contain many references replete with praise for trees and forests. They eulogise the benevolence of forests. One of the Buddhist Sutras says: 'The forest is a peculiar organism of unlimited kindness and benevolence that makes no demand for its own sustenance and generously extends the products of its life and activity affording protection to all beings'. Unfortunately, rapacious eyes of some highly unscrupulous people don't even spare those benevolent and life-sustaining forests, resulting in massive decimation of the same over time. Debased money grubbers, opportunists and grossly avaricious people come together and form a vile nexus to carry out illegal logging in order to make quick money and amass huge wealth. Ironically, lax monitoring, lower conviction rate, complicity of regulatory authorities and lack of stringent rules put together tend to offer ample opportunities of illegal logging in forests which include harvesting, transporting, processing, buying or selling of timber in violation of national laws. Illegal logging further extends to harvesting wood from protected areas, exporting threatened plant/tree species and falsifying official documents. This mercenary nexus may also engage in breaking licence agreements, tax evasion, corrupting government officials and interfering with access and rights to forest areas.

This multi-billion dollar flourishing industry runs and thrives across all continents, from Brazil to Canada, Bolivia to Vietnam and Colombia to Malaysia. It further extends to Cameroon, Kenya, Indonesia, Russia and many more, decimating nature and wildlife, damaging communities and distorting international trade.

It is difficult to estimate and arrive at a consensus figure for the extent of illegal logging worldwide. However, the World Bank estimates that governments worldwide lose $10-15 billion each year as a result of illegal logging. By buying timber and timber products with 'no questions asked, consumer countries knowingly or unknowingly give financial incentives to those committing forest crimes and in the process undermine the efforts to enforce the law in some of the world's poorest timber-producing countries. The illegal logging industry blossoms because of surging international demand for timber, paper, packaging stuff and derivative products. In Indonesia it is largely resorted to for plantation of oilpalm. It is not confined to underdeveloped countries alone but encompasses rich countries of Canada and Russia. The current economic and regulatory environment in most geographies make the clandestine chopping of trees far more lucrative and easy to carry out than doing it the legal way.

Illegal logging, which is often run by large international mafia, occurs in tandem with other crimes such as drugs and human trafficking. In some countries, activists and journalists who speak out against illegal logging are met with threats, violence, and even murder. In 2012, a forest activist, Chut Wutty, was killed while investigating illegal logging in Cambodia. Months later, a journalist covering illegal logging in the country was found dead in the trunk of his car, likely due to an axe wound to the head. Sometimes, doubts are raised on the very credentials of genuine NGOs taking up cudgels against illegal forest destruction.

A Whopping Share of Illegal Logging

Indicative estimates of Wood Resources International, European Forest Institute, Seneca Creek Associates *et al* impart a startling picture of the extent of illegal felling/logging in some select countries. Countries with high percentage of illegal logging include Cambodia, Bolivia, Peru, Indonesia, Ecuador, Gabon and Papua New Guinea. Countries like Russia have different percentages in different regions; its Far East (RFE) reports up to 50 per cent of illegal logging whereas in North-West it is 10-15 per cent. Similarly, Brazil has indicative logging varying between 20-47 per cent whereas Vietnam reports 20-40 per cent of illegal logging. An estimated 10–15 billion euro is lost through illegal logging globally each year. The Graphic indicates the extent of illegal logging in different countries.

Graphic 10.1: Countries with Indicative Illegal Logging [High percentage]

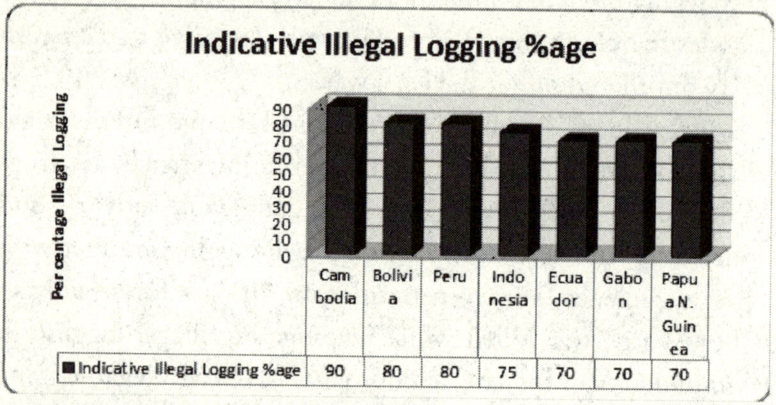

*Data Sources: SavcorIndufor Oy, Seneca Creek Associates, Wood Resources International, FAO, European Forest Institute et al.

Graphic 10.2: Countries with Indicative Illegal Logging [Medium per centage]

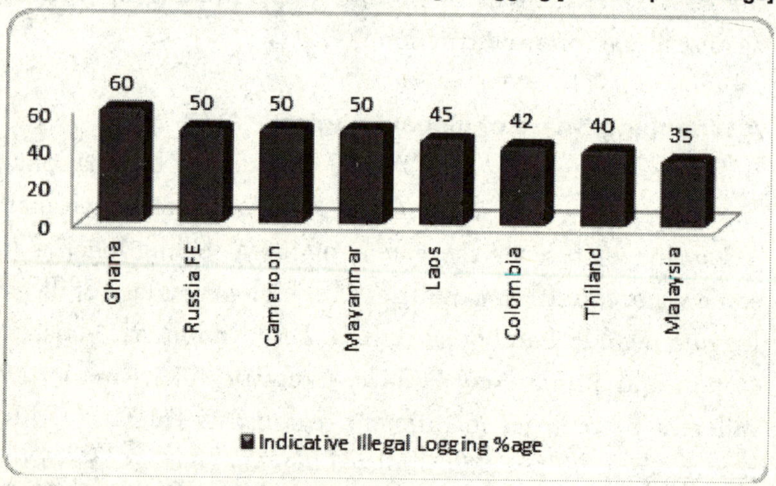

*Data Sources: SavcorIndufor Oy, Seneca Creek Associates, Wood Resources International, FAO, European Forest Institute et al.

Let us now analyse a few select countries where the menace of illegal logging is more pronounced. We shall also delve into the nature and magnitude of the problem along with modus operandi undertaken by wood launderers/loggers in different geographies.

Flourishing Felling in Russian Far East

Siberia in Russia is evocative of remoteness, wilderness, stunning vistas, wild rivers, secret military bases, bears, Amur tigers, untracked areas with no trails and many more things one may crave for! If one wishes to relish massive swathes of unpopulated wilderness, remote and wild shores, then Russia is perhaps the best place to be. Temperature extremes with mercury dipping to –68 C in some places or shooting up to +45 degrees at other places make the terrain highly unevenly distributed in terms of population. In some areas there are no people, no villages...no signs of life!

Forests cover about half of Russia's land mass. Eastern Siberian Taiga forests cover more than 25 per cent of Russian land. The area is divided by the two mighty river systems of Yenisey and Lena. Taiga forests are full of needle-leaf evergreen conifers. These jungles are replete with many varieties of trees such as spruce, durian pine and fir. In the RFE, two categories of forests dominate: Ussuri broadleaf and South Sakhalin-Kurile mixed forests. They cover a vast area of over 2.10 lakh sq km.

Russian President Vladimir Putin once admitted that Russia, the world's biggest exporter of logs, is having its timber stolen at an unprecedented rate. Corruption is ubiquitous. Complicity of government officials, elected representatives, persons occupying statutory high positions and mercenary wood contractors flourishes. Hardwood cut in the endless forests often ends up as flooring and furniture in the glittering markets of United States, Europe, Japan and China. At a meeting on timber management, Putin stated that illegal logging has increased by nearly 70 per cent over the past five years and timber thieves have no problem selling their product.

Illegal loggers and contractors in Russia resort to several tactics. Sometimes they burn up a mammoth area of forest to conceal over-cutting. If a violent storm hits any area and the licence is given to clean out uprooted trees, they carry away much more than stipulated quantities, they cut much more than allowed, or cut species that aren't permitted. The wood mafia is often linked to violent organised crime and they tend to undermine the whole administrative apparatus by paying off local officials. The forests of the RFE are virtually being pushed to the brink of destruction due to pervasive, large-scale illegal logging, chiefly heading to Chinese manufacturers and subsequently to the US and EU in finished form.

One report reveals that in 2010, the volume of Mongolian oak logged for export was twice the amount legally authorised for harvest from the region. At least half of oak shipped across the border to China was stolen.

Environmental Investigation Agency (EIA), a US-based NGO, recently exposed two companies Xingja (China) and Lumber Liquidators (USA). As per their investigation, the Chinese company was reportedly involved in illegal logging in the forests of RFE and the US company was regularly obtaining furniture and floorings made out of this stolen wood in contravention of Lacey's Law. Following the EIA investigative report, federal agents searched Lumber Liquidators headquarters and one of its stores in Virginia. The Lacey Act amended in 2008 stipulates US companies conduct rigorous due diligence to exclude illegally sourced wood and imposes legal liability for the import of illegal wood products.

Illegal logging is now increasingly being perceived as a great threat in the RFE, which provides an important habitat, both in terms of shelter and food, for unique animals such as the Amur tiger. Only about 450 of these beautiful animals are left in the wild. More commonly called the Siberian tiger, it preys on deer and wild boar, which in turn live on acorns and walnuts that grow in one of Russia's most diverse forests. Both oak and walnut trees in the forests are targets of illegal loggers. Oak and walnut wood are highly prized for flooring and furniture and much of it is taken by companies with valid logging permits.

Illegal logging is also putting in peril the livelihoods of thousands of villagers and indigenous peoples inhabiting the pristine forests. WWF-Russia forest policy projects coordinator Nikolay Shmatko concluded about the state of Russian illegal logging: 'With minimal efforts in place to detect and prosecute illegal logging throughout the region, the sheer scale of violations has reached epidemic proportions.'

Peru

Peru in west-central South America is bordered on the north by Ecuador and Colombia, on the east by Brazil and Bolivia, on the south by Chile and on the west by the Pacific Ocean. The forest region called Selva is the largest and most sparsely populated in the country. The altitude ranges from 3,600 m down to 700 m. The broad Amazon plain is the least populated zone, not only of the forest region, but also of the whole country, so that its original features are still relatively intact.

Peruvian forest's importance stems from the sheer fact that it is home to the second largest portion of Amazon rainforest and the third largest expanse of tropical forests on the planet, after Brazil and the Democratic Republic of Congo (DRC). This richly forested country is one of the Top 10 biodiversity rich countries globally and boasts of nearly three-quarters of the world's tropical glaciers.

In 2012, the NGO EIA published a report that detailed the systemically corrupt, appalling and at times 'magic' process by which Peruvian wood is exported to the US. EIA's report put stress on the social conditions involved in the logging—migrant workers 'kept almost as hostages', self-perpetuating debt traps, 'sexual slavery practices', narco-traffickers and organised crime connections—as well as strategies to confuse the authorities, including 'lists of trees that do not exist in the real world' and discs of round-wood 'planted' in the ground to resemble real tree stumps.

Almost 70 per cent of all the logging concessions have been suspected of 'major violations', and almost 50 per cent have been cancelled 'due to confirmed major violations.' This means that out of the total 609 concessions in the entire Peruvian Amazon—36 per cent of which have not been inspected—almost 45 per cent are either under investigation or have been cancelled. Illegal logging threatens all forested lands, including protected areas and indigenous territories. EIA identified over 112 shipments of illegally logged cedar and big-leaf mahogany from Peru to the US between January 2008 and May 2010. The volume represented 35 per cent of all cedar and mahogany shipments from Peru to the US. In fact, large-scale industrial concessions give rise to perpetuate large-scale illegal wood laundering. It is ironical that illegal timber extraction takes place using the same regulatory documents designed to ensure sustainable logging. Peru badly needs to rectify the entire concession system and upgrade forest monitoring apparatus as well to prevent further decimation from recalcitrant loggers.

Indonesia

Illegal logging here has destroyed some 10 million ha of forest. Indonesia's wood-processing industries operate in a strange legal twilight, in which major companies that—until the economic crisis of 1997—attracted billions of dollars in Western investment, obtain more than half their wood supplies from illegal sources. Wood is routinely smuggled piercing the borders to neighbouring countries, costing heavily the Indonesian government in terms of revenue loss year after year. As a direct consequence of widespread deforestation, Indonesia has now become one of the largest emitters of greenhouse gases, mainly due to the conversion of its forests and carbon-rich peat-lands. These shifts in land use have ecological and social consequences, as Indonesia's forests are home to thousands of plant and animal species, and 50-60 million Indonesians depend directly on the forests for their livelihoods.

Every year, Indonesia loses over 6 lakh ha of rainforest, endangering species like Sumatran tigers and orangutans. As few as 400 Sumatran tigers are left in Indonesia's forests as palm oil plantation rapidly destroys their habitat.

Indonesia's rainforests and carbon-rich peat-lands are being destroyed to make the disposable consumer products invariably found in our shops: paper for our glossy magazines, toilet paper, packaging and palm oil for products like toothpaste and chocolate.

This destruction of forests also impacts the wider world; peat-lands are perhaps the world's most critical carbon stores, and Indonesia's peat-lands are vast, storing around 35 bt of carbon. When these peat-lands are drained, burned and replaced by palm plantations, carbon dioxide is released and conditions are set for devastating forest fires, which were responsible for Singapore's 'haze wave' in 2013. The destruction of peat-lands is much more devastating than simple forests because when one hectare of peat-land is drained for oil palm production, about 3,750-5,400 tonnes of carbon dioxide is released over 25 years, whereas the carbon dioxide released from clearing a hectare of tropical forest, without peat bogs, is 500-900 tonnes.

No doubt, Indonesia can't abandon benefits of oil palm because once planted, this tropical tree can produce fruit for more than three decades, providing much-needed employment and livelihood for poor rural communities. The plant yields more oil per hectare than any major oilseed crop. Indonesia is now the leading supplier for a global market that demands more of the tree's versatile oil for cooking, cosmetics and bio-fuel. But oil palm plantations often replace tropical forests, killing endangered species, uprooting local communities, and contributing to the release of greenhouse gases.

Millions of hectares of former forest are now covered in degraded forest remnants, scrub and the ubiquitous alang-alang grass. With this loss of forest, Indonesia is losing biodiversity, wood supply, income and ecosystem services.

Indonesia announced a moratorium in 2011 on the award

of new licences in primary natural forests and peat-lands. This is the most ambitious conservation scheme ever launched in the country. However, the country lost some 8,400 sq km of forests in 2012 alone and the pace of losses remains unabated. That year, annual primary forest loss in Indonesia was estimated to be even higher than in Brazil where clearance of the Amazon basin has usually accounted for the biggest losses. Norway pledged $1 billion to Jakarta if it slows forest losses.

Bolivia

Sometimes referred to as the Tibet of the Americas, Bolivia is a beautiful, geographically diverse, multi-ethnic country in the heart of South America. It shares with Peru control of famous Lake Titicaca, the world's highest navigable lake.

Bolivia is the twelfth most biodiverse country on Earth, home to at least over 2,000 known species of amphibians, birds, mammals, reptiles, and more than 17,000 species of plants. In 2005, Bolivia's biodiversity made science news headlines when Dr Robert Wallace of the Wildlife Conservation Society discovered a new species of titi monkey in the Madidi protected area. The monkey is now known for its distinctive territorial song.

Forests encompass nearly 53 per cent of Bolivia's total geographic area. Bolivia has substantial rainforest cover in its lowland areas: the Bolivian Amazon covers about 59.6 mha, of which roughly two-thirds is forested. About half of Bolivia's forest cover consists of primary forest. The average annual rate of deforestation between 2005 and 2010 was 0.5 per cent. Ten per cent of forests are under the management of private companies. During the 1990s, government granted 20 mha to timber companies, while large swathes of forest were cleared for soybean and coca cultivation. Though the government passed laws that required the logging industry to replant forests to ensure sustainability, loopholes made it possible for many firms to bypass the requirement. Further, illegal logging operations smuggled timber into Brazil where it was exported as Brazilian wood. Greater threats to Bolivia's forests emanate from oil and gas development, commercial agricultural expansion, subsistence agriculture and fuel-wood collection, and land-clearing for cattle pasture. In 2005, fires set for land-clearing turned out of control during the drought period and around 5 lakh hectares of forest and pasture land went up in smoke. Agricultural fires are likely to worsen in the future as population pressures mount and the Amazon region experiences drier conditions due to Climate Change.

Illegal logging is quite ruthless in the area of Chore and Amboro. Forests are cleared giving way to coca plantation. Logging at Amboro Park continues. Timber pirates find ways to escape the vigilance of the few and poorly equipped rangers, continuing their onslaught against a natural wealth of inestimable value.

A study by Fundación Amigos de la Naturaleza (FAN), one of the most credible environmental institutions in South America, reveals Bolivia lost nearly 1,820,000 ha. Deforestation was mainly prompted by the expansion of mechanised farming, followed by livestock rearing and small-scale agriculture. In early 2013, the government passed a law requiring all landowners guilty of illegal logging prior to 2011 to reforest the cleared area, or ensure its use for productive agriculture, in exchange for reduced fines. Bolivia is also a partner country of the UN-REDD Programme and the Amazon Cooperation

Treaty Organisation (ACTO) and has worked with the other seven member-countries to develop strategies aimed at curbing illegal logging in the Amazon.

Brazil

Over half of the Amazon rainforest is located in Brazil. Nearly two-thirds of Brazil's land is covered by forests, of which 92 per cent are primary forests, one of the highest percentages in the world. Thirty-seven per cent of Brazil's forest areas are managed by local and indigenous communities, and 17 per cent are designated as protected areas. It is also the country with profuse biodiversity housing more than 56,000 described species of plants, 1,700 species of birds, 695 amphibians, 578 mammals and 651 reptiles.

The bulk of Brazil's forest cover is found in the Amazon Basin, a mosaic of ecosystems and vegetation types including rainforests, seasonal forests, deciduous forests, flooded forests, and savanna. This region has experienced an exceptional extent of forest loss over the past two generations—an area almost certainly exceeding 600,000 sq km, or about 15 per cent of its total surface area. The increase in Amazon deforestation in the early 1970s coincided with the construction of the Trans-Amazonian Highway, which opened large forest areas to development by settlers and commercial interests. In recent times, growing populations in the Amazon basin coupled with increased viability of agricultural operations, have given a fillip to deforestation rates. Deforestation in Brazil has turned out to be one of the most important global l environmental issues today.

There is significant evidence that illegal logging is quite widespread in Brazil. In recent years, Brazilian environmental enforcement agency Ibama made several large-scale seizures of illegally harvested timber including one in September 2003 when 17 people were arrested for allegedly cutting 10,000 hectares worth of timber. Logging in the Amazon is closely linked with road building. One study conducted by the NGO Environmental Defense Fund shows that areas that have been selectively logged are eight times more likely to be settled and cleared by shifting cultivators than untouched rainforests because of access granted by logging roads. Logging roads give colonists access to rainforest, which they exploit for fuel-wood, game, building material, and temporary agricultural lands.

Greenpeace in May 2014 found rampant and proliferating illegal felling and wood laundering thriving in Brazil. Timber goes to buyers in the UK, US, Europe and China. Several forest management officials are reportedly implicated in the nefarious business.

With little oversight, big landowners obtain permission to cut down more trees than they intend to log and then sell unused credits to lumber mills and other farmers. According to the NGO Imazon, the situation is rapidly getting worse by the fact that the area illegally logged increased by 151 per cent in Para and by 63 per cent in Mato Grosso in 2011-2012. In April 2014, the Brazilian Environment Ministry said fraud in Para was responsible for the unlawful sale of 26.8 million cum of forest products. It also admitted that the control system used by the state is flawed. In Santarem, one of the biggest cities in Para, a local environment ministry chief is alleged to have colluded with logging firms.

The government has accorded much attention in recent years to the impact of agribusiness in Brazil, including soy and livestock production, on forest resources. The government was instrumental in the establishment of the Amazon Fund in 2008, a voluntary mechanism by which developed countries offer financial compensation to developing countries that have achieved significant reductions in their rate of deforestation. Government efforts have rendered encouraging results; deforestation fell by 70 per cent in the Brazilian Amazon region from a ten-year average of 19,500 square km per year in 2005 to 5,800 sq km in 2013.

Wild-forest fires are regular phenomenon in Amazon Basin of which Brazil is also a part. The 2019 Amazon rainforest wildfires season saw a year-to-year surge in fires occurring in the Amazon rainforest.

Fires normally occur around the dry season as slash-and-burn methods are used to clear the forest to make way for agriculture, livestock, logging and mining, leading to deforestation of the Amazon rainforest. Such activity is generally illegal within these nations but enforcement of environmental protection can be lax. The increased rates of fire counts in 2019 led to international concern about the fate of the Amazon rainforest, which is the world's largest terrestrial carbon dioxide sink and plays a significant role in mitigating global warming.

The increasing rates were first reported by Brazil's National Institute for Space Research (Instituto Nacional de Pesquisas Espaciais, INPE) in June-July 2019 through satellite monitoring systems, but international attention was drawn to the situation by August 2019 when NASA corroborated INPE's findings and smoke from the fires, visible in satellite imagery, darkened the city of São Paulo despite being thousands of kilometres from the Amazon. By the end of August 2019, INPE reported more than 80,000 fires across all of Brazil, a 77 per cent year-on-year increase for the same tracking period.

Cambodia

During the Vietnam conflict in the late 1960s, a good many forests were severely degraded. Deforestation continues. As of 2016, only 87,424 sq km forest remained, including 28,612 sq km of evergreen forest. Between 2001 and 2014 the country lost a total of 1.44 mha of forest with the current annual rate estimated to be 14.4 per cent. Forests occupy more than half (57 per cent) of land expanse in Cambodia. Nearly a third of these forests are within protected areas, although just 3 per cent comprise primary forests.

For the majority of Cambodia's people, life is often short and tough. Nearly 70 per cent of the population subsists on less than $2 a day. Despite this, Cambodia is rich in natural resources particularly timber, minerals and petroleum. The government has leased 45 per cent of the country's land to private investors.

In the aftermath of Cambodia's civil war, both the Khmer Rouge (1975-79) and the Phnom Penh government used logging to fund military campaigns. The war ended in 1998, but the destruction of the country's forests through illegal felling and concomitant corruption continues. Local lore says when a tree falls in Cambodia's forests, it will likely be cut into three-metre logs, loaded onto a truck, smuggled across the Vietnamese border and then sold for thousands of dollars in a country with a voracious appetite for the type of luxury wood that is becoming increasingly sought after in Cambodia.

The country's renowned environmental activist, Chut

Wutty, was shot dead in April 2012 during an altercation with military police and security officials working for a logging firm in Koh Kong province. Environmental groups report that the destruction of Cambodia's forests has since accelerated.

More than 7 per cent of Cambodia's forest cover has been eliminated, worse than the situation in Malaysia and Indonesia.

Many communities across the country have banded together to try to protect the forests through regular patrols, but their efforts to alert authorities often fall on deaf ears. These patrols often resort to burning piles of expensive timber to prevent companies from coming back to collect them. Local authorities often collude with companies to locate, log and sell sought-after wood species that are supposed to be protected under Cambodian law.

Although the government has put a moratorium on granting concessions, massive tracts of land are leased to private companies in the name of agricultural development. They continue to be at the heart of deforestation. Loggers are illegally felling trees outside these areas and bringing the high-end wood inside the boundaries, where it can be sold.

In a quest for seeking overseas land resources, two Vietnamese companies Hoang Anh Gia Lai (HAGL) and the Vietnam Rubber Group (VRG), have leased vast tracts of land for rubber plantations in Laos and Cambodia.

The huge pressure for land to plant rubber is driven by high prices and soaring international demand, especially from China. As the third-largest producer of rubber globally, Vietnam is a key global player, and HAGL and VRG dominate its domestic production. With limits on the land available at home, both companies have turned to neighbouring Cambodia and Laos.

The governments in Cambodia and Laos are allocating large areas of land, ignoring laws designed to protect human rights and the environment.

As per a recent report by Global Witness in May 2013, both companies are involved in clear-felling intact forest within and beyond their concession boundaries, contrary to provisions in law. HAGL is alleged to have contracted a powerful Cambodian tycoon to clear and process timber from its concessions. Member companies of VRG appear to have high-level connections with Cambodian government officials and have partnered with a notorious illegal logging syndicate.

China

China desperately needs all sorts of natural resources, ranging from rare earth metals to crude oil, from natural gas to ore or, coal to anything on the planet. How they could lag behind in accessing and seeking overseas woods? China in its overzealous drive to secure timber, minerals and other natural resources, is increasingly being seen as a predator on the world's forests.

Around over half of global timber shipments are now destined for China. Its companies are importing mammoth amounts of illegally harvested wood. According to Forest Trends, a Washington-based think tank, China consumes more than 400 million cum of timber annually to feed both its burgeoning exports and growing domestic demands. Production of paper products has also grown dramatically in China, doubling from 2002 to 2007. China has now become the biggest global consumer of tropical timber, getting imports from nearby Siberia, Asia-Pacific region and Africa.

A Chatham House report indicated that China's imports of

illegally sourced wood increased dramatically during 2000–04, later declining in 2008 from the peak, largely owing to the reduced supply of illegal logs from Indonesia and Burma. It further established China as the world's largest importer of illegal wood in volume terms and assigning 20 per cent of overall imports to be of illegal origin.

All relevant Chinese government agencies are now engaged in addressing illegal logging and trade, and the government has held formal discussions with both the EU and the US. Lack of relevant legislation has prevented the Chinese authorities from stopping specific shipments of timber entering the country despite requests from source countries and evidence of illegal origin.

Chinese companies and entrepreneurs have also penetrated African countries of Gabon, Ghana, DRC, Mozambique and Cameroon. China's trade relationship with Africa has deepened to the extent that it has become Africa's biggest trading partner. From 2000 to 2014, African trade with China has risen from $10 billion to $220 billion. Latest data of 2019 reveals Chinese investment in Africa to the tune of $2 trillion since the beginning of 2005. China also delivered African Aid Package (AIP) in 2018 to further consolidate its position in Africa.

Investments by Chinese state-owned and private enterprise, mostly in natural resource sectors, are witnessing an increasing trend. Many of these investments are in forested or woodland areas, some are directly in forest management and logging enterprises, and many others use timber and other forest products. In the DRC, there are no official Chinese-owned forestry concessions but an apparently increasing share of the 4 million cum of timber currently produced each year under artisanal permits is being bought by informal Chinese timber traders.

African country Gabon's nickname is 'the country of green gold'. Forests cover more than 85 per cent of the country's land area and account for more than 10 per cent of its exports. There are now around 30 big Chinese timber companies operating in Gabon alone.

China, in other words, now substantially catalyses deforestation across the globe. A report 'Appetite for Destruction: China's Trade in Illegal Timber' says that in 2011, China imported at least 18.5 million cum of illegal logs and sawn timber worth around $3.7 billion. Forty-seven per cent of these illegal logs (5.6 million cum) came from Russia while 21 per cent (2.5 million cum) came from Papua New Guinea. The remaining illegal logs last year came from the Solomon Islands (12 per cent), Myanmar (4 per cent), and DRC (4 per cent), Equatorial Guinea (2 per cent) and Mozambique (1.5 per cent). The report adds, 'More than half of China's current supplies of raw timber material are sourced from countries with a high risk of illegal logging and poor forest governance'. The report also underlines China's shifting strategy: once forests are exhausted in one country, it moves onto another.

Madagascar

Madagascar presents a scene of greed and extreme methods to secure the coveted wood. A government coup in 2009 led to the ravaging and despoiling of Madagascar's vanishing forests, including rainforests within protected areas. Loggers targeted the highly valuable rosewood, despite a ban on cutting the increasingly rare species. EIA found that 95 per cent of the rosewood was at the behest of China's unscrupulous traders. The fashion for reproduction furniture made from rare rosewoods has created an upsurge in illegal logging from the Mekong basin to Madagascar.

Growing Urbanisation

Around half of the world's urban population lives in urban centres, with Latin America and Caribbean Islands topping at 79 per cent. In the annals of global demographic changes, 2009 was clearly as the first year when more people lived in urban spaces than in rural areas. It is believed around 60 per cent of global population would reside in urban areas by 2030.

Burgeoning urban population has been exerting tremendous pressure on forest areas, denuding and depleting their resources. New research now convincingly corroborates strong linkages between urbanisation and pressure on forests. Steady increase in city-dwellers pushes demand for foodstuffs and puts immense pressure on adjoining forests. Since urban populations are expected to continue growing, the pressure on forests could become even more intense in the future. The key drivers of tropical deforestation have shifted from small-scale landholders to domestic and international markets that are distant from the forests.

An extensive analysis of satellite, demographic and economic data suggests strong and significant correlation between urban growth, agricultural exports and deforestation. The two primary reasons: as there are fewer persons farming the land, more mechanised agriculture takes place. In addition to this, as more people move to the city where incomes and rates of consumption are generally higher, more pressure is put on forests to produce more animal and processed food products, which require more clearing. Some non-food agricultural items, like sugarcane or palm oil plantations grown for bio-fuels which are frequently geared for international markets, significantly increase pressures on forest areas, as demand for those products in developed countries grows.

Tackling the Menace of Illegal Logging

Illegal logging is a critical and pressing international issue, one that represents a major concern for buyers of forest products also. Illegal logging undermines good forest governance, threatens biodiversity, increases carbon emissions, and reduces the potential for forests to provide stable supplies of products and support local communities.

In order to combat illegal logging, WWF has created the Global Forest and Trade Network (GFTN), a global partnership comprised of suppliers, producers and purchasers from across the forest industry supply chain who are willing to lead the industry in sustainable forest management and trade. The GFTN strives to eliminate illegal logging by helping its participants overcome forest management and responsible purchasing challenges while progressing towards credible certification. WWF also promotes forest certification through the Forest Stewardship Council (FSC).

In the US and the EU, procurement policies increasingly require proof that forest products are derived from known and legal sources. These include actions such as the amended United States Lacey Act and the European Union Timber Regulation, both of which prohibit the trade in illegally harvested timber and timber products.

The EU's international licensing system for legal timber called the Action Plan on Forest Law Enforcement, Governance and Trade (FLEGT) is another significant measure to thwart illegal logging and wood laundering. The plan is based on voluntary partnership agreements with timber producing countries, which will establish systems designed to identify legal products and license them for import into the EU and unlicensed products that will be denied entry at the EU border. The FLEGT Action Plan also proposes Voluntary Partnership Agreements (VPAs) between the EU and individual timber-producing countries. Legally produced timber exported to the EU would be identified by means of licences issued in partner countries. Timber originating in a FLEGT partner country and arriving at an EU point of import without such a permit would be denied entry.

The Sustainable Forestry Initiative (SFI) is a 'forest certification standard' and programme of SFI Inc., a non-profit organisation. Third-party forest certification programmes like SFI are recognised tools to demonstrate due care/due diligence as they show that a company has taken proactive steps to avoid illegal forest products through its supply chain.They are widely exercised in the US and Canada. The SFI programme takes significant steps to promote legal and responsible forest management and procurement worldwide. The SFI standard requires SFI programme participants to 'comply with applicable federal, provincial, state, and local laws and regulations' and that when buying fibre from offshore avoid controversial sources of supply, including illegal logging and fibre sourced from areas without effective social laws.

Corruption is not confined to forests but spreads far and wide. Systemic corruption of the police, prosecutors and judges ensure those with sufficient resources can literally get away with it. There are no quick fixes for combating corruption and that without an effective criminal justice system; all the current policies remain ineffective. More often than not, those who are caught in the act of smuggling or logging and arrested are the poor or vulnerable. The main culprits behind this crime remain unscathed and roam scot-free. And while civil society plays an important role in monitoring the forests and exposing those behind the crimes, little is being done to properly address the more complex issues that expose corruption.

Following the money is one such method, and coordination between domestic and international enforcement should work. But this takes enormous courage, determination and political will, something that can only occur when anti-corruption activists find their champions within governments and are extended the support needed to ensure those who are getting away with it feel the heat.

Meat consumption has a bearing upon our forests. Rearing animals requires much larger areas of land than agriculture. Nearly one-third of the ice-free surface of the planet is converted into pasture for domestic animals and 30 per cent of available arable land is used to grow livestock feed rather than food for direct consumption. The global demand for meat is constantly rising but space to rear livestock is not. Around 70 per cent of the Amazon rainforest clearing is done to make way for cattle ranches. If you opt to reduce the amount of meat you eat, you will lower the global demand for meat and help prevent further destruction of forests to make way for more livestock.

A layman could also proactively take part in the global efforts to curb this monstrous practice of illegal felling by simply sharing posts, stories, pictures reflecting illegal logging or poaching in jungles on social media. This would certainly go a long way in both galvanising thoughts and combating the menace.

Tempting Sheen of Metals

PRECIOUS METALS have been luring humankind for millennia together and continue to entice even today! Narratives of gold abound through the annals. Craving for gold is inherent in human thinking, whereby anyone might become exceptionally wealthy almost instantly with the help of sheer adventurism and a bit of luck. The Brazilian Gold Rush enticed people to the far-off lands of Minas Gerias at the turn of the eighteenth century.

Minas Gerias, the territory that is now present-day Brazil, was colonised in the fifteenth century with the arrival of Portuguese coloniser Pedro Alvares Cabral. Following the acquisition of the territory, the Portuguese started to exploit natural resources of the region. They harboured the notion of discovering rich gold and silver deposits similar to those found and exploited by the Spanish in their South American colonies. But for the first two centuries of Portugal's occupation of the region, success in obtaining precious metals remained elusive. During the late seventeenth and early eighteenth centuries, *bandeirantes*—groups of adventurers comprising Portuguese colonists, indigenous Indians, cabaclos and people of mixed race, led adventurous expeditions into the interiors of Brazil and began to traverse the rugged mountain chains west of Rio de Janeiro and the southern interior of the colony.

In the latter half of the seventeenth century, gold was finally discovered. The discovery caused such a stir that by the last decade of the century, a considerable proportion of the population of Sao Paulo, Rio de Janeiro and the northern province of Bahia had rushed to the site of the discovery.

At the turn of the eighteenth century, hundreds of thousands of people were agog to set sail for Brazil from Portugal. Such was the rush that the government in Lisbon had to implement legislation to stop the Portuguese from emigrating *en masse*. Though it has not been officially recorded how much gold was extracted from the operations in Brazil, it is believed that, at the height of the gold rush, some 350,000 oz of gold was extracted every year.

Eternal Glitter of Gold

Humans have been using gold as a form of currency or exchange dating back to the earliest civilisations. Gold was a commodity highly prized by the ancient Egyptians, as it was held to be the flesh of the sungod Ra, and was thus considered a symbol of eternal life. Our fascination with the beauty and versatility of this yellow metal spans thousands of years. The earliest record we have of human use of gold dates backs to 5,000 BC. It was in ancient Egypt that humanity's obsession with gold as a symbol of wealth and power was firmly established. It was also used as a fashion accessory in Eastern Europe. It is highly valued because of its unique colour, lustrous shine and malleable properties by which it can be easily moulded into jewellery.

Another notable appearance of gold in human history is in the Middle East in 3,000 BC. A civilisation in modern-day Iraq used gold to create intricate jewellery designs that are still popular today. However, Egyptian cultures have the most extensive history of the use of gold, having begun burying it with Pharaohs for their wealth in the afterlife as far back as records go. They used gold to create statues, to coat armour and as currency. It was the expansive Egyptian culture's use of gold currency that spurred the worldwide adoption of gold as the standard of international trade. Afterwards, gold's popularity across the globe exploded. But the Egyptians continued to be considered the masters of gold craftsmanship for many years after.

Availability of Gold

Gold is a rare commodity and much gold might have buried underground as forgotten treasure. It is difficult to arrive at an exact figure as various estimates differ. Thomson Reuters suggests that there are only 171,300 tonnes of the precious metal in the world. Warren Buffett, one of the world's richest investors, says the total amount of gold in the world (171,000 tonnes)—the gold above ground, that is—could fit into a cube with sides of just 20 metres. Or to put it another way, it would reach 9.8 metres above ground level if exactly covering

Wimbledon Centre Court. This is all about the existing available gold that mankind has extracted so far. So how much is left? Well, the good news is that we are not likely to run out of gold any time soon. The US Geological Survey estimates there are 52,000 tonnes of mineable gold still in the ground and more is likely to be discovered.

WORLD'S LARGEST GOLD MINES

Here's a look at the three largest gold mines in the world:

Grasberg Mine, Indonesia

This biggest gold mine of the world is operated by Arizona-based miner Freeport-McMoRan Copper and Gold. It is also one of the largest copper mines. Grasberg is located in the province of Papua in Indonesia. Open-pit operations at the mine began in 1990 and the company expects that they will continue through 2016. There are still expansion projects in progress in the Grasberg mineral district, including the development of large-scale, high-grade underground ore bodies. The company produced 1,061,000 ounces of gold in 2016.

Yanacocha Mine, Peru

Located in the Andes Mountains in Peru, the Yanacocha mine is the largest gold mine in Latin America. It opened in 1993 and contains three active pits. Since its inception, the mine has produced more than 26 million ounces of gold. Yanacocha is owned and operated by Colorado-based gold producer Newmont Mining. The company holds a 51.35 per cent interest in the property and the rest is claimed by Compania de Minas Buenaventura. In 2012, Yanacocha produced 1.346

million ounces of gold and 534.69 koz (kilo oz) in 2017. As of 31 December 2012, the asset had 3 million attributable ounces of gold reserves. It is part of a joint venture with Newmont's Conga Project in Peru. Together they are known as Minera Yanacocha.

Goldstrike Mine, USA

The Goldstrike mine is located on the Carlin Trend in north-eastern Nevada and is run by Barrick Gold. Mining at Goldstrike began in 1987. The property is a compilation of the Betze-Post open-pit mine and the Meikle and Rodeo underground mines. In 2012, it produced 1.17 million ounces of gold at a total cost of $541 per ounce. In 2015 it produced 642,493 oz of gold which increased in 2018 to 795,663 oz. Goldstrike's proven and probable gold reserves are estimated at 12.3 million ounces.

Top Gold Producing Countries in the World

Gold mining has assumed the form of a global industry over the past few decades. As of now, 90 countries mine this glittering metal. The Top 20 producing nations account for more than 75 per cent of the world's gold each year.

Gold is one of the rarest elements in the world, making up roughly 0.003 parts per million of the earth's crust. In 2018, global gold mine production was reported at 3,332 tonnes. This figure is up 2% from the previous year (2017) and is the largest year-on-year growth in the last four years, according to the GFMS Gold Survey 2019. The driving forces behind increase in production came out of operations in Argentina, the US, Russia and Mali.

Until 2006, South Africa was the top gold producer in the world which was relegated to eighth position in 2018. China holds now top position in gold mining. Below is the list of Top 10 gold producers in 2018.

1. China

China's mining production stood 404 tonnes and gold reserves of 2,000 tonnes in 2018. Zijin Mining Group which reportedly produced 2,770,428 ounces of gold in 2011 is China's biggest producer.

2. Australia

The country's primary gold assets are located in Western Australia, South Australia and New South Wales. About two-thirds of global gold production comes from Western Australian mines, and the entire continent has estimated gold reserves of 9,800 tonnes. The Super Pit is Australia's largest open-cut gold mine. It produces up to 850,000 ounces of the precious metal each year and is operated by Kalgoorlie Consolidated Gold Mines. Australia produced 319 tonnes in 2018, occupying second slot in the world.

3. Russia

Russia held 5,300 tonnes of known gold reserves in 2018, according to the US Geological Survey. The production of gold increased by 6.8 per cent year-on-year in 2012. Production in 2018 stood at 297 tonnes, surpassing the US.

The largest Russian gold company Polyus Gold International increased its gold production by 12 per cent in 2012 compared to the year before, reaching approximately 1.68 million ounces.

4. United States

The United States' gold production stood at 222 tonnes in 2018 and its reserves around 3,000 tonnes. One of the world's top gold producers, US-based Newmont Mining, delivered output of 4.98 million ounces of gold in 2012 through operations in five continents, including two in North America.

5. Canada

It produced 193 tonnes of gold in 2018. Canada inched up two spots on the list and occupied the number five spot in 2018, producing 17 tonnes more year-on-year. In 2017, Toronto-based Seabridge Gold stumbled upon a significant goldfield in British Colombia which is estimated to contain a whopping 780 metric tonnes of gold. This could be source of enhanced production in coming years.

6. Indonesia

Indonesia overtook Peru to grab the number six spot, producing 196 tonnes of gold in 2018. The island nation posted the biggest production increase in 2018 of 23 per cent, or 36 tonnes. A high-grade development in the Grasberg open pit operation drove most of the gains, with the giant producing nearly 40 per cent of the country's total gold output. The Indonesian government has been restructuring environmental and taxation policies, which were negatively impacting the mining industry for a few years.

7. Peru

Gold output fell for a third consecutive year (2016, 2017, 2018) in Peru, by 7 tonnes, largely due to crackdowns on illegal mining operations in the La Pampa region and lower grades at existing projects. It produced 155.4 tonnes in 2018. Mining is a significant portion of Peru's economy and accounts for over 28 per cent of the region's total output.

8. South Africa

Once the top gold producer in the world by a wide margin, South Africa's gold mines have been giving lower returns every year since 2008, with the exception of 2013 when production rose by a few tonnes. Its gross output stood at 123.5 tonnes in 2018. In 2012, South African gold mine strikes caused output to drop. It was the country's worst year of gold production since 1905. In 2012, South African gold output amounted to roughly 6 per cent of total production worldwide.

The nation is struggling with rising costs of electricity and labour, with many mines closing due to unprofitability. South Africa is, however, still home to the world's deepest gold mine, the Mponeng mine, extending 2.5 miles underground.

9. Mexico

Its production stood at 121.6 tonnes, slightly less than South Africa in 2018. Although production fell by 4 per cent and for a third consecutive year, Mexico remains a competitive gold source. Output has risen from just 50.8 tonnes in 2008 to over 130 tonnes in 2017, one of the largest increases in a nine-year span. Mexico is an attractive place for mining due to a relatively low cost of regulation.

10. Ghana

Ghana is a promising country and its production level crossed

100 tonnes attaining 101.8 tonnes in 2018. Ghana is Africa's second largest producer of gold, and could soon beat out South Africa for the top spot, and is also known for its reserves of various industrial minerals. Bullion production rose by just a tenth of a tonne in 2018 and accounts for over 20 per cent of the nation's total exports.

The Great Gold Rushes beginning 1970s

The world witnessed the first great gold rush in 1971. Gold prices rose from $35 to over $700 per ounce, a twenty-fold rise! A decade later, prices settled near $300, nearly a ten-fold increase over the $35 base! The second great gold rush commenced in 2001 and gold prices climbed from $275 to over $675. The third great gold rush began in 2007 when gold prices climbed above $750 that year! The fourth great gold rush occurred from 2008 to 2011 when price of gold increased to approximately $1,900 per ounce. Current gold price stands at $1,673 per ounce.

Popular analogies maintain that when the US dollar goes down in value, investors seek safety for their money and buy gold. Also, when there is global turmoil, gold tends to shine. Apart from this, China has over 1.34 billion people and an appetite for base metals such as copper, zinc, nickel, and iron, along with gold.

For over a half century, the Chinese people were banned from owning gold. But all that changed in 2003, and now the enormous demand is causing an important paradigm shift, as gold and silver move from the West to the East. Another factor in the paradigm shift is official Chinese demand from the People's Bank of China (PBOC) which is diversifying some of its massive foreign exchange reserves, some $3 trillion.

Chinese Fresh Rush for Gold

China National Gold Group Corporation (CNGGC) or China Gold, China's largest gold conglomerate with primary interests in mining and refining, is on the hunt for global acquisitions and partnerships. China is the world's biggest producer, importer and buyer of gold, giving the country increasing sway over prices, output and the global gold market in general. The country's official gold consumption has nearly tripled. This is apparently encouraging overseas acquisitions. China's long-term demand for gold in China will remain strong as the younger generation buys gold online.

Acquisitions by China Gold would revive a mostly moribund market for gold mergers and acquisitions. This has been seen in both the mining and investment segments of the gold market. China Gold's current preference is for assets in countries near China, such as Mongolia, Russia and Central Asia. It also is looking for acquisitions in developed countries such as Canada, Australia and the US. The company is also exploring opportunities in developing countries, including in Africa and South America.

China Gold is also working on potential partnerships with Barrick Gold Corp. of Canada, Newmont Mining Corp. (USA) and Kinross Gold Corp.

If China's largest gold producing company is successful, the alliance would bring one or both of the world's largest western gold mining companies closer to China. It could mean an important new source of supply for the insatiable demand that is coming from China. Zhongyuan Gold Smelter Co Ltd is the largest gold refiner and bar refinery in China and part of CNGGC is a subsidiary of China Gold (Zhongjin Gold Co Ltd), which is headquartered in Beijing.

UK's Hummingbird eyes Overseas Gold

Gold mine developer Hummingbird Resources entered into a conditional agreement to acquire all of Gold Fields' mining and exploration assets in West Africa's second largest gold producer Mali. The all-scrip deal, calculated to be worth $20 million, would make Gold Fields Hummingbird Resources' largest shareholder, with 26.3 per cent of the company's enlarged capital. The prime Gold Fields asset in Mali, likely to be the first mine to be brought onstream by Hummingbird, is the Yanfolila gold project.

Burgeoning Interest in African Gold

Africa accounts for up to 30 per cent of global production, with South Africa one of the global leaders. Zimbabwe has historically been a prominent gold producer in Africa. Now Tanzania and Mali have several mines being commissioned or in final feasibility stages. Ghana is another significant gold producer in Africa.

Gold mine production is centred on underground and open pit operations in Archaean age greenstone belts in Ghana, Tanzania, Burkina Faso, Cote de Ivoire (Ivory Coast) and Zimbabwe. Nearly all of South Africa's production is centred on the Witwatersrand placer gold deposits, which are mined almost exclusively underground. Exploration efforts centred on West Africa as well as East Africa are concentrating on surface weathered gold-bearing lateritic and saprolitic rock, which is very suitable for relatively low cost surface mining methods, including heap-leach gold extraction. Much of the recent and ongoing success in tapping these mineable gold reserves is simply due to the fact that low-grade (to 1.3 grams per tonne gold) near-surface oxidised ores can be recovered economically.

Several of the world's largest mining corporations are active in African gold mining and development. The merged gold producer Barrick Homestake, one of the largest gold producing companies in the world, has commissioned the Bulyanhulu mine in Tanzania, which began production in mid-2001. Initial production estimated at around 300,000 ounces per year, but should increase to full production at about 500,000 ounces per year. This project alone will contribute 10 per cent extra gold to Barrick Homestake's annual production.

AngloGold is active in South Africa, Namibia, Tanzania, Mali and Zimbabwe. Its presence in Mali increased with the go-ahead to develop the Yatela deposit, the company's second gold mine in the country after Sadiola. Gold Fields is active in South Africa and Ghana, with its new open pit operation, Tarkwa. Placer Dome's purchase of 50 per cent of the South Deep project in South Africa is rated as one of the world's largest undeveloped gold mines, with 52 million ounces (1474.17 tonne) reserves.

Harmony Gold Mining Company acquired Randfontein Estates Gold Mines as well as Elandskraal Gold Mines in South Africa. Harmony has holdings in three countries, primarily in South Africa, Australia and a single producer in Canada, and operates a refinery in South Africa. Harmony and ARM have also acquired AngloGold's Free State assets.

Avgold is primarily a South African producer with some of its operations producing from outside of the traditional

Witwatersrand basin in the Barberton greenstone belt.

Ashanti is one of Africa's leading gold mining and exploration companies. It has four operational mines in Ghana and facilities in Zimbabwe, Guinea and Tanzania. Rio Tinto Zimbabwe produces 10 per cent of Zimbabwe's annual gold production through its two gold operations and dump retreatment operations.

The Six Metals of PGMs

The importance of Platinum Group Metals (PGMs) can be gauged by the fact that one in four of the goods manufactured today either contain PGMs or had PGMs play a key role in their manufacture. Platinum, palladium, iridium, rhodium, ruthenium and osmium—these six metals are indispensable to our daily lives. They find applications ranging from fountain pens to aircraft turbines, from anti-cancer drugs to mobile phones, from catalytic converters to ceramic glazes.

PGMs mostly occur together in natural deposits alongside nickel and copper. Mineable deposits of PGMs are very rare, with annual production amounting to around 400 tonnes. Rhodium, ruthenium, iridium and osmium are mined as by-products of platinum and palladium. The industry is capital and labour intensive. Companies are now investing heavily in exploration and production facilities to extract more of these precious and rare metals. Platinum and palladium are the most important metals of the PGMs owing to their economic value, extensive usage and higher quantities. Here, we shall describe them at length.

Platinum, the White Gold

Which element is rarer and scarcer than gold? What element is a better conductor than silver and denser than lead? The answer is platinum–30 times scarcer than gold, the strategic metal for modern industries and armies. An inquisition into shooting platinum demand takes us to the Eastern hemisphere and China's rapidly expanding economy. In particular, the growing Chinese middle class is generating massive demand for new automobiles, which in turn is consuming plenty of platinum. For the past one decade, auto-catalysts have constituted 40-50 per cent of total global platinum demand. Auto-catalysts utilise platinum to clean the emissions of motor vehicles, and 95 per cent of the world's new passenger cars now come equipped with them. Both auto production and emissions standards are steadily increasing around the world, especially in the huge emerging market of India, Indonesia, Malaysia and China. South Africa produces 80 per cent of the world's platinum.

Element platinum has roots in the Spanish word *platina,* which means 'little silver.' The chemical symbol of platinum is Pt and the atomic number is 78. It has six naturally occurring isotopes and is an extremely rare element. Platinum is also the least reactive metal. It is also considered to be a noble metal because of its strong resistance to corrosion and high temperatures.

Generally, it is found in a chemically uncombined state as native platinum. It can be found naturally in the alluvial sands of some rivers, so it was once used by the natives of South America to create artefacts. Platinum has various uses: it is used in catalytic converters, laboratory equipment, electrodes, thermometers, watches, spectacle frames, dentistry equipment, defence equipment and jewellery. It is one of the most valuable metals in the world since only a few hundred tonnes are produced each year. There are health concerns with platinum since it is a heavy metal, but it is not as toxic as other metals. Platinum is considered to be among the most visually appealing metals. It has a silver white colour, has a great lustre, and is soft and malleable. It is the most ductile out of the pure metals. It also does not oxidise, regardless of the temperature. However, it can be corroded by halogens, cyanides and sulphur. Since platinum is a very durable and not easily susceptible to wear and tarnishing, it is often used to make jewellery. It does not corrode very easily, withstands high temperatures, and has good electrical properties. Because of these properties, it is also often used in industrial applications.

Platinum is one of the rarest metals. It only occurs at a concentration of 0.005 ppm in the Earth's crust and is mistaken for silver sometimes. Native platinum is alloyed with iridium as platiniridium, so it is often found chemically uncombined. Native platinum is more commonly found in secondary deposits, as in platinum combined with other metals. Platinum is found commercially as a by-product from the processing of nickel and copper mining.

As new applications continue to emerge, so the demand for platinum grows. Platinum's use is growing in fibre optics and medicine. Platinum in certain compounds can inhibit the growth of cancerous cells.

In 1990 international platinum prices were averaged at \$471.29 per ounce. As of March 2020, they stand at \$896 per ounce. Platinum witnessed tremendous surge in the year

2008 when it touched an all time high of $2,273 per ounce and average prices stood $1,571 that year. For the period 2006-2015, prices remained firm and stood above $1,000 each year. Historically (1986-2020), platinum prices were highest in the year 2011 which recorded average prices $1,719.03 per ounce.

Flourishing Platinum Mining in South Africa

The history of platinum mining can be traced as far back as ancient Egypt, where platinum—probably without knowledge of its eventual value—was used for writing inlays on little statuettes. Platinum mining would only be introduced to western civilisation in the seventeenth century—when Spanish conquerors considered it a waste product of gold mining in Columbia. That is when these conquerors named the mysterious grey-white metal *platina*.

Today, the automotive industry requires the powerful catalytic properties of platinum for use in exhaust systems as catalytic converters, which reduce harmful effects of gasses such as carbon monoxide into less harmful carbon dioxide and water vapour. These various and increasingly important uses of platinum have led modern demand for platinum mining in South Africa to outstrip supply.

Platinum can take up to six months to be refined. On an average, 7-12 tonnes ore yield one ounce of high grade platinum. The beneficiation process begins with the underground extraction of platinum-rich ore, which is then disintegrated into workable chunk sizes. The froth flotation method is used to extract the metal by mixing these particles with reagents, and having air pumped through the material. Platinum-containing particles float to the top. The skimmed-off material is smelted at temperatures exceeding 1,500°C, which enables the separation of the platinum metal from waste products.

South Africa possesses over 80 per cent of the world's platinum group metal reserves. Along with Russia, platinum mining in South Africa produces a total of 90 per cent of the world's platinum demand—which is about 130 tonnes per year. The Merensky Reef, stretching from southern Zimbabwe through to the Rustenburg and Pretoria regions, is the centre of platinum mining in South Africa, playing host to companies like Rustenburg Platinum Mines and Bafokeng Rasimone Platinum Mines.

Anglo American Platinum Ltd (AMPLATS) is the industry leader in the mining, marketing, and distribution of this precious mineral as well as other platinum group metals, AMPLATS produces 40 per cent of the world's total platinum group metals. Other key platinum mining companies in South Africa include the likes of BHP Billiton and Impala Platinum. The establishment of the Twickenham Expansion Project, 100 km south-east of Polokwane, will see the production of 250,000 t/m pure platinum.

These expansions are set against the already existing large-scale platinum mining in South Africa, such as the well underway Impala Platinum No. 20 Shaft Project, which is geared at producing 185,000 ounces of platinum on the Bushveld Complex, which has been mined since 1897.

Significant Platinum Producing Countries & Companies

In 2018, the world's total production of platinum was 160 metric tonnes, with South Africa being the leading producer of the metal, according to a report published by the US Geological Survey (USGS).

However, the year 2018 saw a decline in platinum production compared with 199 metric tonnes in 2017, due to several factors such as low prices for the metal in the market, workforce reduction, and mine-shaft closures.

Top 5 countries with the highest platinum production in 2018 are described below:

South Africa

South Africa's reserves of platinum group metals (PGMs) stands at 63,000 metric tonnes. It is the world's largest producer of platinum, processing 110 metric tonnes in 2018. Owing to various factors such as job losses and mine-shaft closures, the country witnessed a sharp fall in production in 2018 compared to the 143 metric tonnes recovered in 2017. In August 2018, Impala Platinum, one of the leading platinum miners in the world, announced plans to close five shafts at the Impala Rustenburg mine. A significant decline in the US dollar platinum price along with sustained high mining-cost inflation was cited as the reason for the restructuring.

South Africa is home to the Bushveld Complex, which is considered to be the largest resource for PGMs in the world. Some of the prominent platinum producing mines in the country are the Impala Rustenburg mine, Mogalakwena mine, Marikana mine, Bathopele mine, Khomanani mine, and others.

Russia

Russia is the second-largest producer of platinum in the world, with a 2018 production of 21 metric tonne. A bulk of Russia's production of PGMs comes from Norilsk Nickel's operations, which are primarily based on the Taymyr Peninsula in East

Siberia, and those located on the Kola Peninsula.

The Norilsk mining centre in the Russian High Arctic is the main producer of the commodity in the country. Owned by Norilsk Nickel (Nornickel), the mining centre is considered to be the world's largest producer of palladium.

In 2018, Nornickel entered into a 50:50 joint venture partnership with Russian Platinum to develop disseminated ore deposits located in the Norilsk Industrial District. The JV has a target of producing nearly 70-100 tonnes of PGMs annually from Nornickel's Maslovskoe deposit and Russian Platinum's South flank of Norilsk-1 and Bokoni Chernogorskoe deposits.

Zimbabwe

Zimbabwe stands third among the top platinum producing countries in the world, with a production of 14 metric tonnes in 2018. The African country recorded similar numbers for its platinum production in 2017. Zimbabwe's reserves of PGMs are 1,200 metric tonnes.

The country's major platinum mines are the Mimosa, Ngezi and Unki mines. The Mimosa mine is located in the Midlands province and is owned by Mimosa Investments, a JV between Implats and Aquarius Platinum. The Ngezi mine is one of the largest resources of platinum in Southern Africa. This Zimbabwean platinum mine is owned by Zimplats and Impala Platinum.

The Unki mine is located in central Zimbabwe in the Midlands Province. It is owned by Anglo American Platinum.

Canada

Canada takes the fourth spot in the list of global platinum

recovery, with a production of 9.5 metric tonnes in 2018. A significant part of platinum mined in the North American country is a by-product of nickel mining. The country's reserves of PGMs are 310 metric tonnes. Canada exports the majority of its output to the US.

Ontario's Sudbury Basin hosts a majority of the country's platinum producing mines. Raglan nickel mine in Quebec and the Lac des Iles mine in Western Ontario are the other platinum-producing sites. The country's River Valley in Alberta is the largest undeveloped primary PGM resource.

In June 2019, Sibanye-Stillwater signed an acquisition agreement with Generation Mining (Gen Mining) for the development of the PGM-copper Marathon project in Canada. The deposit is located 10 km north of the town of Marathon, Ontario, and on the eastern margin of the Coldwell Complex, which is a proterozoic layered intrusion.

The proposed project includes three open pits, an ore processing plant, tailings and mine rock storage facilities, and a 7 km power transmission line. The mine will have a production capacity of approximately 22,000 tonnes per day.

In August 2019, Canada-based mining company North American Palladium (NAP) announced the acquisition of a 51 per cent stake in the Sunday Lake Project from Impala Platinum (Implats). The project is located 25 km north of Thunder Bay in Jacques Township, and approximately 60 km south of NAP's Lac des Iles Mine operations (LDI).

USA

The US is the fifth largest platinum producer in the world, with a production of 3.6 metric tonnes in 2018. The country owns two platinum producing mines including the Stillwater Mine, located in Nye, Montana, and the East Boulder Mine, which is located by Big Timber, Montana.

The Stillwater platinum/palladium mine is in the Beartooth Mountains, 130 km southwest of Billings, Montana. Production at the Stillwater mine commenced in 1986. Located 20 km west of Stillwater, the East Boulder project began commercial production in 2002.

In December 2016, South Africa's Sibanye Gold agreed to acquire the Stillwater Mining Company, which owns the Stillwater Mine and East Boulder Mine, for $2.2 billion.

The current mineral reserve is estimated to sustain the Stillwater West section until 2044 and that of the East Boulder mine until 2056.

The production of platinum metal is highly concentrated, with the five largest companies Anglo American Platinum, Impala Platinum, Sibanye-Stillwater , Lonmin plc and OJSC MMC Norilsk. Top 6 companies and their production figures of 2018 are depicted below:

1. Anglo Platinum: 1,292,000 oz

The assets of Anglo American Platinum Limited (Amplats) comprise of 11 managed mines, as well as a number of JVs and associate mines across South Africa and in Zimbabwe. Most of the ore from these mines is processed at one of Amplats' fourteen own concentrators before being smelted at one of the company's three refineries in South Africa.

2. Impala Platinum: 1,016,000 oz

Impala Platinum's (Implats) operations are focused around

the Bushveld Complex in South Africa and the Great Dyke in Zimbabwe. The company's primary operational unit is on the western limb of the Complex near Rustenburg. Implats also owns a 73 per cent stake in Marula on the eastern limb. In Zimbabwe, the company operates Zimplats and has an interest in Mimosa Platinum. Impala Refining Services (IRS) smelts and refines the metal from the company's own ore as well as material purchased from third party companies.

3. Sibanye-Stillwater: 819,000 oz

Sibanye-Stillwater is a leading international precious metals mining company, with a diverse portfolio of platinum group metal (PGM) operations in the United States and Southern Africa, gold operations and projects in South Africa, and copper, gold and PGM exploration properties in North and South America.

This company is one of the world's largest producers of platinum, the second largest primary producer of palladium and a top tier gold producer, ranking third globally, on a gold-equivalent basis, as well as a significant producer of rhodium and other PGMs and associated minerals such as chrome.

4. Lonmin plc: 679,000 oz

Lonmin was initially incorporated as the London and Rhodesian Mining and Land Company Ltd (Lonrho) in 1909. The company's primary operation—the Marikana mine—is located on the western limb of the Bushveld complex.

5. OJSC MMC-Norilsk Nickel: 653,000 oz

Norilsk Nickel (Nornickel) is one of the world's largest

producers of nickel (accounting for 17 per cent of global nickel production) and palladium (41 per cent of global production). It is also a major producer of copper, gold, silver, iridium, selenium, ruthenium and tellurium. Headquartered in Moscow, Nornickel operates in the Taimyr and Kola Peninsulas (both in Russia) as well as in mines in Botswana and South Africa. As Russia's largest mining company, Nornickel also extracts and refines cobalt, silver, gold, tellurium, and selenium as by-products.

6. Northam Platinum: 307,000 oz

Northam is an integrated PGM producer with operations in the Bushveld Complex in South Africa. The company's primary facility is the wholly-owned Zondereinde platinum mine and metallurgical complex, which extracts roughly 300,000 oz of PGM concentrate annually. Toll refining for PGM concentrate takes place under contract with WC Heraeus (Germany), and is delivered on a weekly basis to Heraeus' Hanau facility where platinum, palladium, rhodium, gold, silver, ruthenium and iridium are separated. Working with Heraeus, Northam also refines and markets PGMs extracted from Platmin's Pilansberg Mine.

Palladium-Auto Boom Fuelling the Demand

Palladium is a metal close to platinum in chemical composition, with the chemical symbol Pd and atomic number 46. Like platinum, it has a white-silver colour and is a very rare metal. It was discovered in 1803 by William Hyde Wollaston. The metal is named after the asteroid Pallas. Palladium belongs to a group of elements called the platinum group.

Palladium is the least dense and melts at the lowest temperature of all of these metals. About half the supply of palladium is used for catalytic converters. These are used to convert toxic gases from auto exhaust to substances that are less harmful. It is also used in the production of electronics, in dental equipment, medicine, chemical applications, groundwater treatment and jewelry. Palladium is important in the technology that is used for fuel cells, which create electricity, heat and water by combining hydrogen and oxygen.

Most of the deposits of palladium are found in South Africa, Montana, Ontario and Russia. It is also extracted from used catalytic converters. It has a lot of investment interest because it is used in various applications but there is limited supply.

Palladium resembles platinum in that they have the same colour and lustre. However, it is not as resistant or durable as platinum. When annealed, it becomes soft and docile, but when cold, it has a much higher strength and hardness. It also dissolves slowly in sulfuric, nitric and hydrochloric acid. It does not tarnish when exposed to air because it does not react to oxygen at normal temperatures. It will tarnish in a moist atmosphere that contains sulphur. Platinum has seven isotopes, which include six table isotopes.

Russia was the top producer of palladium in the year 2007 with a 44 per cent share. Other countries that followed include South Africa, Canada and the US, which are the only significant producers of palladium. As of now, South Africa and Russia combined account for close to 80 per cent of global supply of palladium. It is found alloyed with gold and other platinum group metals in various deposits in places like

Australia, Ethiopia, and North and South America. However, these deposits have a very minor role in the production of palladium. Commercially, the most important deposits are found in Ontario and Siberia.

Palladium can also be extracted from spent nuclear fuel in nuclear fission reactors, but there is no equipment to extract palladium from radioactive waste yet. Palladium is used in a variety of applications. It is used in catalysis, where it is used to speed up hydrogenation and de-hydrogenation reactions. It is used in electronics as well, in multilayer ceramic capacitors. It is also used often in jewelry. Palladium is used in developing photographs as well.

Global Palladium Production and Consumers

Around two-thirds of global palladium use was attributed to catalytic converter production in 2010. Other areas where palladium finds application are electronics, jewellery, medical purposes, dentistry, hydrogen purification, groundwater treatment and other industrial applications. Palladium is a key component of fuel cells, which are used to convert chemical energy from fuel into electricity. About 16 per cent of global palladium use was accounted for by electrical engineering

Country	Production of Palladium in 2017 (metric tons)
Russia	81
South Africa	78
Canada	19
USA	13
Zimbabwe	12
Rest of the World	8.4

Sorce: Worldatlas

application in 2010. Russia and South Africa were the leading palladium producing countries in 2017 followed by Canada and the United States.

Palladium Future Scenario

Palladium has been enjoying a boom, with prices doubling in three years (2017-2019). The metal has seen increasing demand and prices due to its usage in catalytic converters in cars, which are becoming mandatory as the automotive industry looks to clean up its environmental performance. Figures support this growth: global production reached 214 tonnes in 2017, compared to annual gold production of around 175 tonnes, and the metal saw its price skyrocket from $562.98 per ounce at the end of 2015 to $1,261.78 per ounce three years later, an increase of 124 per cent.

This growth has also encouraged a number of new players to step onto the global palladium stage, with miners of all sizes suddenly interested in the metal. With the majority of the world's palladium produced by a single company—Russia-based Norilsk Nickel, which was responsible for 41 per cent of global production in 2017—other miners are looking to develop new projects to challenge this monopoly.

New and resurgent projects

The surge in demand for palladium has seen all manner of miners flock to the metal. Canadian miner Ivanhoe has completed a feasibility study at its Platreef mine in South Africa, a project in which it owns a 64 per cent stake, and aims to produce 219,000 ounces of platinum a year. Similarly, the Lac Des Illes mine in Canada restarted production in 2010 after a two-year hiatus due to resurgent prices. Its owner, North American Palladium, plans to continue underground operations until 2027, and produce 2.32 million ounces of the metal over the next eight years.

However, the former project is not exclusively a palladium operation, while the latter boasts a short lifespan and a history of uncertain finances; regarding new, specialised palladium mines, US-based Platinum Group Metals is one of the companies investing the heaviest in new palladium operations.

"The Waterberg mine would be one of the largest low cost [platinum group metals] mines in the world, with palladium as the dominant metal," said PGM CEO Michael Jones, when asked about his company's new Waterberg mine in South Africa.

The numbers certainly support this claim; a PGM feasibility study, published in September 2019, noted that 63 per cent of the mine's total resources were palladium, and that there would be 19.5 million ounces easily accessible as reserves. The project will cost around $874 million to build and PGM expects to produce around 420,000 ounces of palladium a year, close to double that of Ivanhoe's Platreef project. It has the financial support of industry heavyweights Implats and JOGMEC, which have purchased stakes in the mine of 15 per cent and 13 per cent respectively.

Copper—Asia Fuelling the Demand

Copper is one of the first metals ever to be extracted and used by man. Copper perhaps first came into use as the earliest non-precious metal employed by the Sumerians and Chaldeans of Mesopotamia (present-day Iraq) 5,000-6,000 years ago. These early people developed considerable skill in fabricating copper. The rudiments of craftsmanship then spread to the river-dwelling people of Egypt, where it continued to flourish for thousands of years. Early man found out how to make copper into tools around 5,500 BC and by combining with tin to make bronze, it was even more useful around 3,000 BC. Copper is easy to mould, shape, stretch and is resistant to corrosion. It is a conductor of heat and electricity and used in building construction, power generation and transmissions, electronic products, industrial machinery, copper wiring and plumbing materials.

In the automotive field, copper is extensively used in motors, wiring, radiators, connectors, brakes and bearings. A small vehicle has around 44 pounds of copper in its construction and as much as 100 pounds go into hybrids. Anti-microbial properties of copper reduce transfer of disease and germs, circuitry, microprocessors and electric motors. The copper-nickel alloy is used in hulls of ships as it doesn't corrode in salt water; it reduces drag and increases fuel efficiency because sea life like barnacles don't adhere to it.

Copper is 'green' as it is the most widely recycled metal worldwide—about one-third of all copper consumed is recycled. It is easily re-melted to use directly or reprocessed into refined copper without losing any valuable properties. No less than 160 minerals are derived from copper, the most common being chalcopyrite, malachite, azurite and turquoise. Copper is in fact the leader of the pack when it comes to base metals.

Since 1900, demand for refined copper increased from less than 500,000 tonnes to over 20 million metric tonnes in 2012. Currently, global copper reserves are estimated at 830 metric tonnes (USGS 2019 Survey) and annual demand is 28 metric tonnes. By 2050 it could rise tenfold!

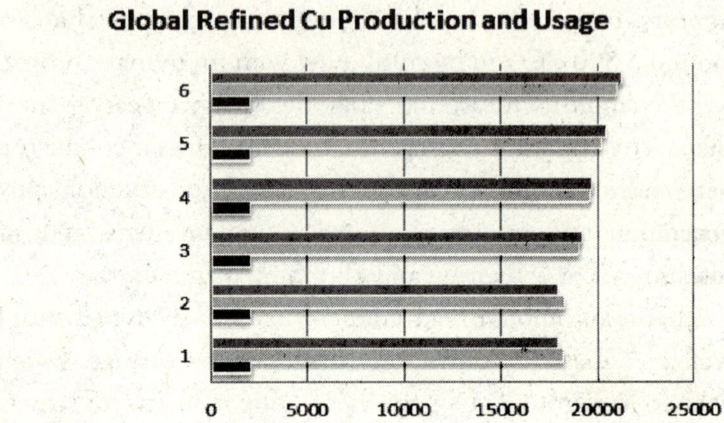

	1	2	3	4	5	6
Refined Usage (TMT)	17888	17894	19129	19697	20390	21278
Refined Copper Production (TMT)	18214	18249	18985	19598	20136	20973
Year	2008	2009	2010	2011	2012	2013

*Data Source: ICSG (International Copper Study Group)

** TMT- Thousand Metric Tonnes

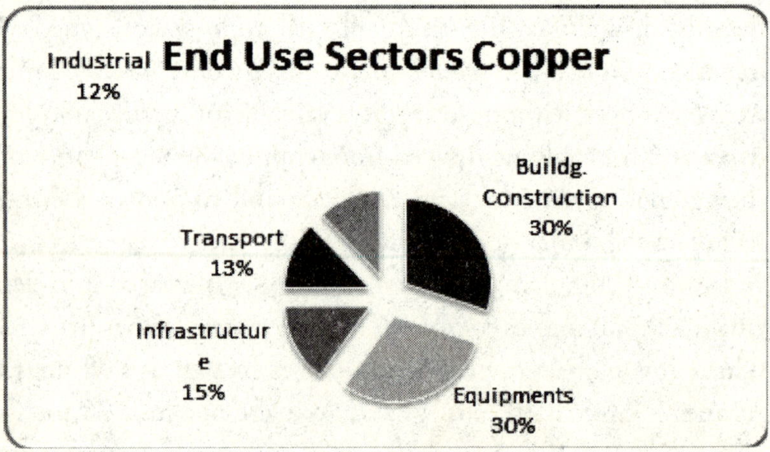

*Data Source: IWCC (International Wrought Copper Council) and ICA (International Copper Association)

Increasing Use of Copper in Transportation

Copper is now being used extensively in new generation airplanes and trains. New high-speed trains could utilise nearly double the 1-2 tonnes used in traditional electric trains. Today's average mid-size automobile contains about 22-25 kg of copper whereas luxury cars may contain around 1,500 copper wire measuring around 1.6 km in length. Automobiles and trucks rely on copper motors, wiring, radiators, connectors, brakes and bearings.

Chile accounted for over one-third of world copper mine production in 2019 with mine output of over 5.6 metric tonne. Other prominent countries are Peru, China, USA, Russia, DR Congo, Australia, Zambia, Mexico, Kazakhstan, and Indonesia in decreasing order.

Sectors and Regions Fuelling Copper Demand

Building construction and equipment are the two prime sectors spurring copper demand which together constitute 60 per cent end-users of refined copper. Others are infrastructure, transport and industrial sectors. The Graphic defines end-user sectors.

Around 60 per cent of the total copper production is consumed in Asia. Next comes Europe at 20 per cent and the US 14 per cent. The rest of the world together consumes merely 6 per cent.

China's eye on Peruvian Copper

China is a big producer as well as seeker of copper. In recent years, the demand for mineral products has gradually increased with rapid economic growth. China has now become the largest consumer of non-ferrous metals such as copper, aluminum and nickel in the world, of which it has relatively low reserves. China Minmetals Corps inked a deal worth $5.85 billion for Glencore Xstrata's Las Bambas copper mine in Peru in April 2014. This is the largest outbound metal mining purchase by any Chinese company. Las Banbas is the largest copper mine under construction and could produce 450,000 tonnes of copper concentrate in the first five years after it reaches full capacity.

China previously acquired Northern Peru Copper at $450 million through China Minmetals Corp and Jiangxi Copper Corporation. Similarly, Aluminum Corporation of China acquired Toromocho Copper Mine of Peru Copper Inc. at the price of $790 million.

Both Latin America and Africa are now at the centre stage of resource extraction. Africa is estimated to hold around 30

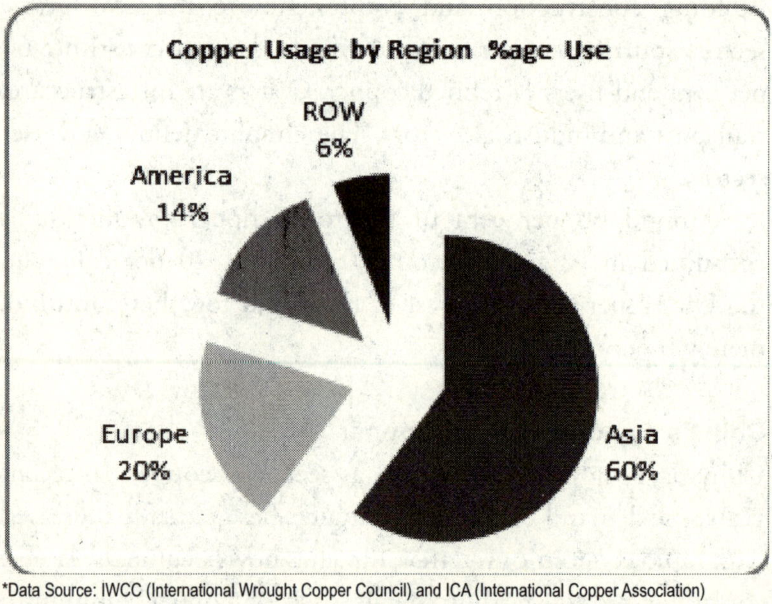

*Data Source: IWCC (International Wrought Copper Council) and ICA (International Copper Association)

per cent of world's mineral reserves and produces more than 60 different types of minerals, metals and ores. Besides traditional western extractors and developers, many new players from the emerging world have joined the fray. Prominent among them are China, India, South Korea, Brazil, Turkey, Thailand, Singapore, Malaysia and Indonesia.

The irony is that a large majority of African populace still reel under abject poverty. While according land and extraction rights to foreign countries, land rights and food security of native dwellers are often bypassed. In many instances, land deals are not transparent and greedy companies acquire huge tracts of land at heavily discounted rates in connivance with local authorities, inviting the wrath of local inhabitants and civil society groups. Proprietary environmental standards are often overlooked while resorting to mining of metals, ores and other resources. The mining and extraction sector woefully lacks sound environmental impact assessment, water use and contamination studies, consensual rehabilitation plans, etc. This is leading to instability, rising conflicts and severe undermining of the ambient environment.

Resources Limited Contenders Uninhibited

For too long, the world has been focused on short term growth and development at the expense of our long term survival as we have depleted our Natural Resources at historically reckless rates!

—Gisele Bundchen

Unless a fundamental change drives natural resource use away from the status quo, this use will continue to grow to 190 billion tons and over 18 tons per capita by 2060.

—UNEP-IRP Global Resources Outlook 2019

ALL NON-RENEWABLE natural resources on the planet are finite. Since the dawn of this century, resource extraction activity has significantly heightened.

From 1970 to 2017, the annual global extraction of materials grew from 27 bt to 92 bt, while the annual average material demand grew from 7 tonnes to over 12 tonnes per capita (Assessing Global Resources Use, UNEP 2019).

Hitherto untouched virgin geographies/countries are now witnessing hectic timber logging and mineral extraction. The global race for the resources is neither confined to BRICS economies of Brazil, Russia, India, China and South Africa nor restricted to advanced economies. Many other contenders like Indonesia, Vietnam, Mexico, Argentina, Poland, Taiwan, Thailand, Egypt, Colombia, Turkey and Malaysia are now in the fray.

Let's first have a look at the Top 10 countries with the most natural resources and mining activity in the world.

1. China

China possesses many domestic natural resources and even stronger appetite to control overseas resources. China is one of the leading producers of phosphates, vanadium, tungsten, antimony, graphite, coal, tin, molybdenum, lead, zinc and gold. It is the second top producer of manganese, bauxite, cobalt, silver, and copper. China is one of the top mining countries in the world and its companies hunt for mining contracts in other countries.

2. Saudi Arabia

Saudi Arabia has over 34.4 trillion worth of commodities, the primary natural resource being crude oil. It has the second biggest oil reserves on the planet which account for over 20 per cent of the oil reserves globally. Petroleum was discovered in Saudi Arabia on 3 March 1938, and since then it has been the leading oil exporter. The country controls the sixth largest natural gas reserves. Other natural resources in Saudi Arabia include feldspar, phosphate, sulphur, lead, tungsten, manganese, copper, zinc, silver and gold.

3. Canada

Canada has over $33.2 trillion worth of natural resources including industrial mineral (gypsum, potash, limestone and rock salt), energy minerals (uranium and coal), metals (nickel, zinc, copper, and lead) and precious metals (platinum, silver, and gold). Canada has the third biggest oil deposits right after Venezuela and Saudi Arabia and the second biggest uranium supply. Canada is the leading supplier of phosphate and natural gas. Canada is the third biggest exporter of timber in the world.

4. India

The mining sector in India contributes about 2.5 per cent of the country's GDP and about 11 per cent of the industrial sector's GDP. The mining and metal industry of India was estimated to be worth over $106.4 billion in 2010. India has the fourth largest coal reserves on earth and significant reserves of limestone, petroleum, diamonds, natural gas, chromite, titanium ore, and bauxite. The country accounts for over 12 per cent of the world's thorium production and over 60 per cent of global mica production. India is the leading producer of manganese ore.

5. Russia

The Russian mining industry is one of the biggest in the world, and it accounts for a large percentage of the Russian commonwealth production of a wide range of commodities including mineral fuels, industrial minerals, and metals. Russia is one of the leading producers of vanadium, silicon, palladium, nitrogen, magnesium metal and compounds, copper, arsenic, cement, and aluminum. The natural resources in the country are estimated to be worth over $75 trillion. Russia is the second largest exporter of rare earth metals.

6. Brazil

Natural resources in Brazil are estimated to be worth over $21.8 trillion, some of the top commodities being uranium, gold, iron and oil. Mining in the country is centred on bauxite, iron, tin, copper and gold extraction. Brazil has the biggest uranium and gold deposits, and they are the second biggest iron producers. Even though they have vast oil deposits, the most valuable commodity in the country is timber. Brazil supplies over 12.3 per cent of the world's timber.

7. United States

The mining industry was quite active during the colonial era, but it became an important industry in the US during the nineteenth century after more minerals were discovered. The total metals and coal reserves in the country were valued at $109.6 billion in 2015. The United States has been known as the leading producers of coal for decades now, and currently, they control over 31 per cent of the global coal reserves and a significant amount of timber. The total natural resources in the country are valued at $45 trillion with over 89 per cent of them being timber and coal. The US has a sizeable deposit of copper, gold, oil and natural gas.

8. Venezuela

The South American state is the leading producer and exporter of numerous minerals including oil, iron ore, gold, coal, and bauxite. Venezuela has over $14.3 trillion worth of unused natural resources, and a majority of their minerals are state-controlled. Venezuela has the largest oil reserves which are bigger than the reserves in the United States, Mexico and Canada combined. Venezuela is the third largest Latin American coal producer, right behind Brazil and Colombia. They have the eighth largest natural gas reserves in the world which account for about 2.7 per cent of the world's supply. Venezuela has the world's second-largest gold deposits.

9. DRC (Democratic Republic of the Congo)

One of the most significant sources of export income for the Democratic Republic of the Congo is the mining industry. As of 2009, the country had over $24 trillion in unexploited mineral deposits, including the biggest coltan (Columbite niobium and tantalum elements) reserve in the world. The state is known for its large gold, tin, tantalum, diamond, copper and cobalt reserves. The American Geological Survey estimates that the country has over a million tonnes of lithium. There were over 25 international mining firms in the DRC in 2011.

10. Australia

Mining is the primary industry in Australia and the main contributor to its economy, earning the country over $19.9 trillion per annum. Australia is one of the leaders in uranium and gold mining. Australia has the largest gold reserves in the world, enough to supply about 15 per cent of the world's demand. The country also contributes over 46 per cent of the world's uranium. Australia is known for its vast reserve of iron ore, copper, timber, nickel, oil shale, rare earth metals and coal. Australia is the top producer of opal and aluminum.

Largest Extractors of Material in World

As per the latest UNEP-IRP Report 2018, the countries with

largest Domestic Extraction (DE) per capita are Australia, Canada, China, Russia, USA, Brazil, Turkey, Vietnam, Indonesia and India.

When it comes to DE of materials, China occupies the top slot followed by India. The US is the third largest domestic extractor of materials.

Figure 12.1: Graph Showing Top 10 Countries with ton per capita domestic extraction

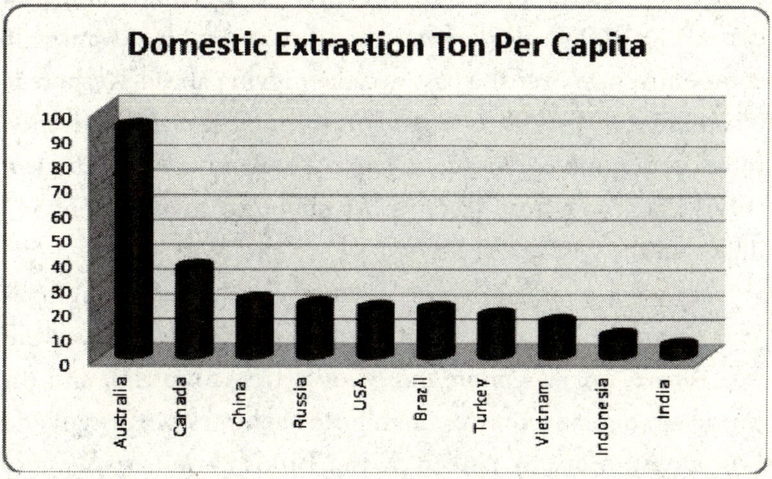

*Data Source UNEP and IRP 2018

There are many other significant contenders raring to have their rightful share and place in the ongoing global race. At this stage, though some might be in evolving stage of cubs and colts while others dark horses but who knows, tomorrow they may become roaring regional tigers. We shall take up the case of AMBIT (Argentina, Mexico, Brazil, Indonesia and Turkey) and analyse their gambit in the global scene. We shall delve deep into their strengths as well as obstacles and bottlenecks.

TURKEY: A High Beta Economy

Turkey is one of the formidable contenders which is likely to occupy a respectable place in the Top 15 economic powers of the world by 2030. Its larger part, also called Anatolia, is in western Asia and the smaller part sometimes referred to as East Thrace in south-eastern Europe.

Turkey is now the eighteenth largest economy in the world with a GDP of $786 billion. In less than a decade, per capita income in the country has nearly tripled and now exceeds $10,000. Turkey has made considerable advances in competitiveness over the past decade, moving ahead 16 spots to No. 43 in the World Economic Forum's Global Competitiveness Index, for example. During this time, Foreign Direct Investment (FDI) has grown from just over $1 billion to an average of $13 billion in the past five years.

Turkey is strategically located and plays an important role in the transit of oil. It is situated at the crossroads between oil-rich former Soviet Union and Middle East countries, and the European demand centres. Significant volumes of Russian and Caspian oil move by tanker via the Turkish Straits to Western markets, hence the Turkish government derives significant revenue from transit fees. In addition, it is home to one of the world's busiest chokepoints, the Turkish Straits, through which some 3 million barrels per day flowed in 2013. Turkey has two domestic crude oil pipelines and two major international oil pipelines to meet demand in Turkey and to transport exports. With a rapidly expanding young and urban population, both consumption and demand for energy are expected to increase. Turkey's proximity to a number of new gas sources, such as Iraq and Azerbaijan, and its location as host to nearly all proposed gas pipelines between the EU, the Middle East and Central Asia, make it an attractive market for power generation. The government has liberalised the electricity sector. These changes have caught the attention of several utility companies.

ARGENTINA: Remarkable Potential to Reinvent and Reach Dizzy Heights

Argentina is now one of the largest economies in Latin America.

Experts are now talking about a new group called CIVETS (Colombia, Indonesia, Vietnam, Egypt, Turkey and South Africa). They propound that, the past decade was all about the BRIC, the massive economies of Brazil, Russia, India and China, which kicked off at the beginning of the new century, boomed and are now slowing.

As of now. only ten countries are responsible for 68 per cent of global extraction and many more countries are vying for a lion's share. The lust for resources is not going to die down easily! A fundamental change therefore is direly warranted in how natural resources are used around the world.

The UN has set high ambitions through Sustainable Development Goals (SDGs) and CBD for global prosperity, the protection of our biological diversity and land resources and limiting global warming. Global GDP has grown four-fold since 1970 against the two-fold increase in global population. These trends have required large amounts of natural resources to fuel economic development and the associated improvements in human well-being. However, these gains have come at a tremendous cost to our natural environment, ultimately impacting human well-being and exacerbating inequalities within and between countries.

Lust, Lure and Consequent Upheaval

IN THE preceding 12 chapters of this book, we learnt about the gradually heating up race for natural resources, avaricious eyes on precious metals and minerals, intense vying for economic supremacy, growing quest for overseas lands and resources. In reality, the race is increasingly becoming fiercer every passing day. Where will this race end? When will it culminate? What will be the final denouements of this global contest? What are the upshots of conflicting interests of nations? What are the consequences of over extraction and consumption of natural resources? Let's try to find answers of each one of these queries.

The early years of this century saw both the relentless quest for natural resources and the most broad-based spell of wealth creation in recent history. The period was remarkable for the breadth of its geographic coverage, with emerging market economies, especially China and India, not just sharing in the growth, but driving much of the action. The story of wealth creation' was broad in its coverage of assets, with both financial assets and non-financial assets growing at a fast pace. More importantly, it was socially inclusive: all levels of society shared in the rewards. The global mean wealth per adult grew at 8 per cent per year between 2000 and 2007.

A report titled 'Riding a Wave of Growth-Global Wealth 2014' by a Boston-based consulting group reveals many interesting facts about burgeoning income patterns. China seems the greatest beneficiary for all of its skyrocketing enterprises. China's millionaire households increased to 2.4 million from 1.5 million in 2012. India sprang up to fifteenth position by recording 175,000 millionaire households in 2013. According to a TOI report, in 2020, India has 5,986 ultra-high net worth individuals (UHNWI), each with over US$ 30 million (around Rs 215 crore), and it is globally ranked 12th on the list of countries with most such people.

Asia-Pacific region has emerged the third wealthiest region in the world owing to strong GDP growth over the past five years in the emerging economies of China, India and Indonesia.

Current use of resources has reached around 60 bt per year. Seeing the trend, it is likely to increase further to 100 bt by 2030. Humans on this globe now extract and use about 50 per cent more than consumed 30 years ago. Almost half of global resource extraction is taking place in Asia, followed by North America with almost 20 per cent and Europe and Latin America with 13 per cent each. Large variations exist, though, in natural resource extraction per capita: On average, an inhabitant of Australia extracts around 10 times more resources than inhabitants of Asia or Africa. Increasing resource extraction is now a leading cause of growing environmental degradation and social problems, more acute in poor countries of Africa, Latin America and Asia.

Seeing the current trend of the mad race for natural resources, it is highly likely that the global extraction and consumption of natural resources will continue to increase dramatically, unless measures are implemented to reduce the overall amounts of resource use. This growth will mainly be driven by increasing consumption in the emerging and developing countries. These countries are now seeking a life-style similar to that enjoyed in richer parts of the world. Many non-renewable raw materials have already reached a peak of extraction, or are about to reach a peak in the near future. Growing global demand for resources will therefore exacerbate rivalry and competition along with engendering possibilities of mild conflicts.

Environmental Degradation due to Relentless Extraction

Around 25 million people the world over are constantly engaged in extracting metal ores, minerals, rare earths, sand, coal and diamonds from the bosom of the Earth through mining activity. This 24x7 activity is wreaking havoc with widespread impacts. It is leading to catastrophic pollution of water bodies in mining areas. Storage of tailings and waste is turning out to be a more damaging activity than the mining operation itself. Vast swathes of land now stand degraded with considerable modification/alteration in landscape. All mining pockets report enhanced air pollution, and people suffer from a number of lung diseases with breathing impairment. Heightened artisanal and small-scale mining are resulting in the silting up of rivers, exploitation of bush meat, deforestation, and mercury and cyanide pollution. Growing pollution of water, land and soils as a direct consequence of mining operations, further leads to bad health of the inhabitants, lost agricultural productivity and damaged ecosystems. Conflicts between mining companies and indigenous peoples over cultural or local resources are surfacing more than ever before and creating social unrest and disharmony. Much of the impact has become irreversible now.

International trade in raw materials and products has also increased manifold in recent decades. Trade in natural resources can support economic development, as it enables resource-rich countries to export resources and raise revenues. If there is adherence to high environmental and social standards, trade can contribute to sustainable development of poor countries. However, this isn't exactly happening as growing world trade has been posing severe environmental and social threats. Current heightened global trade is accelerating resource extraction by linking local resources in all parts of the world with global demand. In addition, the current trade system is giving rise to unequal levels of resource consumption by shifting resources from poorer low consuming countries to richer, high consuming countries.

Accelerating pace of resource extraction requires goods to be transported between the countries of extraction and processing and finishing and finally to countries of consumption. This leads to substantial augmentation in greenhouse gas emissions resulting from increased transport activities. Out of the total energy used in world transport, 95 per cent is still supplied by petroleum. Around a quarter of global energy-related greenhouse gas emissions originate from transport activities.

Swelling trade volumes also contribute to increased land requirements for transport infrastructure, such as roads, harbours and airports besides a sizeable transport fleet.

Global trade of natural resources allows countries with high purchasing power to bolster resource consumption beyond their own national resource capacities. In recent years, more and more countries have become net-importers of natural resources and products and thus consuming more than would be possible based on domestic resources only. Thus many countries are treading on the path of ecological trade deficit (ETD). It may not be construed as problematic that countries with poor natural endowments have net resource imports. However, the inequalities in resource use facilitated by the current patterns of world trade raise concern as they may jeopardise sustainable and equitable development in all world regions.

In order to ensure material welfare for all people, trade could help redistribute resources from countries with a high extraction to countries with lower extraction. Currently, however, the opposite is happening. Europe's per capita extraction is now higher than the rest of the world. It also has the highest net imports of natural resources, at almost 3 tonnes per capita per year. Developing and emerging economies are net exporters of natural resources. Hence, the currently evolving international trade is not balancing—rather, it is fortifying—inequalities in per capita resource use.

China's Belt & Road Initiative (BRI): Enormous Pressure on Global Resources & Environment

The ambitious plan unveiled by Chinese President Xi Jinping in 2013, to boost connectivity in over 70 countries. The world's largest ever infrastructure project includes the financing and building of everything from roads to airports and is estimated to cost $4-8 trillion. However, environmental design requirements are not integrated in the project cycle—especially in early-stage project planning—as a consequence of which Southeast Asians are sure to pay dearly in years to come. It is certainly going to put enormous pressure on global resources to meet the additional demand of aggregates, cement and iron emanating from the mammoth project. Critics call the BRI 'the riskiest environmental project in history'.

A study titled Greening the Belt and Road Initiative released by the World Wildlife Fund (WWF) and HSBC in November 2019 found that apart from the lack of information on sustainable and green investment opportunities, getting the BRI to 'go green' has not gained the proper attention of the financial fraternity nor the wider private sector. Rich in biodiversity, Southeast Asia's fragile ecosystems and the many communities which depend on them for survival are especially at risk due to the BRI's ever-expanding list of infrastructure projects.

Biodiversity at Risk

Chinese-backed hydropower projects along the Mekong River —which spans Cambodia, Lao PDR, Myanmar, Thailand and Vietnam—have seen dams cause river flow changes and block fish migration, leading to a loss of livelihood for communities there which live off the river.

Fish stocks have declined in recent years due to hydropower dams built upstream in Cambodia and neighbouring countries, according to the Fisheries Action Coalition Team (FACT).

Apart from the loss of flora and fauna, deforestation in

areas such as the Pan Borneo Highway—which spans Malaysia, Indonesia and Brunei—also causes landslides, floods and other disasters. Failure to protect the environment means that local communities will have to find other ways to provide ecosystem services such as clean water and air.

The WWF has listed over 1,700 critical biodiversity spots and 265 threatened species that will be adversely affected by the BRI.

This is already happening in Indonesia, where the habitat of the world's rarest great ape—the Tapanuli orangutan—is at risk from a $1.6 billion hydroelectric power plant currently being built in Sumatra's Batang Toru forest highlands. One of Indonesia's most bio-diverse regions, the Batang Toru forest, is also home to the critically endangered Sumatran tiger and Sunda pangolin.

Not only do they have to deal with the loss of habitat, Southeast Asia's native species are also under threat from the increased connectivity the BRI has created.

A study published in the *Current Biology* journal in December 2019 found that the BRI could potentially introduce more than 800 alien invasive species—including 98 amphibians, 177 reptiles, 391 birds and 150 mammals—into several countries along its many routes and developments, threatening their ecosystems.

With some species capable of wreaking havoc worth billions of dollars on a nation's agricultural exports, the study titled Risks of Biological Invasion on the Belt and Road presents a chilling side-effect of the BRI's network of roads, railways and pipelines. Environmental guidelines such as the Ministry of Environmental Protection of China's (MEP) Guidance on

Building the Green Belt and Road are non-binding and hardly implemented in BRI projects.

And as there is no definitive list of BRI projects published by the Chinese government, its impact on the environment will not be confirmed for years to come.

It is undeniable that the BRI, as with any other large development projects, has the potential to cause irreversible damage to the environment. China should mitigate this by integrating environmental aspects into the planning, building and operating of its infrastructure.

Spurt in Overseas Land Acquisition Inflicts Umpteen Woes

Overseas land acquisitions have been rising and the trend is accelerating further. One aspect of this trend is large-scale land deals in which governments and influential companies buy or lease vast acreage of farmland abroad, particularly in Africa and Southeast Asia. Thereafter, this land is utilised to grow food or bio-fuel crops (with tilt towards cash crops) instead of buying these crops on world markets. These deals are causing sharp rise in food prices and increasing concerns over long-term food security. Some say that such deals might increase economic growth, government revenues and opportunities for economic development in the target countries. Contrarily, they are catalysing food shortages and posing a threat to food security, displacing large numbers of local populace and fomenting severe socio-political unrest. These deals are becoming one of the reasons for local land prices shooting up.

Leaderships the world over draw their sustenance from economic well-being and GDP growth and no world leader would like to jeopardise his position by indulging in war. Even the US isn't in that privileged position. But people at the helm of affairs in China seem to harbour the thought of full-scale military intervention to fulfil their misadventures in South-China Sea or elsewhere.

Augmented thirst for extraction and consumption is causing serious damage. Present extraction and resource harnessing patterns need to be changed drastically in order to continue to thrive on this planet. Instead of the headlong rush for natural resources, they should be utilised in a more sustainable manner so that we are able to protect our natural resource base and the fragile ecosystems on this lovely planet.

Global atmospheric levels of CO_2 breached 415 ppm on 11 May 2019, recorded at Mauna Loa observatory, Hawaii. Such elevated levels have never been witnessed since the dawn of human civilisation. Extraction of fossil fuels and use thereof has direct bearing on atmospheric CO_2. Earth's safe carrying capacity is 350 PPM only which was breached in May 1986! The psychological 400 ppm mark too was decisively broken in September 2016. In order to get stabilised at the current level, we need to drastically cut down the use of fossil fuels by 80 per cent. Still, to revert back to 350 levels, it may take decades!

The belligerent sky is now full of fuming greenhouse gases, rendering our planet ever hotter and hotter! Soon, we might find it resembling other solar systems plancts whose horrific details of simmering gases, acidic fumes and molten rocks we often read!

Securing a flourishing future cannot be achieved by continued extraction and use of fossil fuels. Now is the time for concrete action, review and follow-up. It's time for people to come together in small and large groups, in public and private places to multiply our efforts to reduce extraction and consumption patterns and stabilise the rising, life-destabilising gases in the atmosphere.

We, the citizens of this planet, are facing a twin crisis of heightened extraction and global warming. The crisis has been further exacerbated by the unprecedented phenomenon of Climate Change. It's both crisis and emergency, because we lack a firm commitment and resolve to drastically change things. Humanity is hurtling toward a future that strips away the freedom of families to flourish across generations. People's self—imposed discipline of frugal consumption and strong political will on the part of governments across the planet can indubitably avert impending doom!

Abbreviations

IAEA	International Atomic Energy Agency
bcf	billion cubic feet
boe	barrels of oil equivalent
bpd	barrels per day
bt	billion tonne
CDW	Construction and Demolition Waste
CTBT	Comprehensive Test Ban Treaty
cum	cubic metres
DE	Domestic Extraction
DRC	Democratic Republic of Congo, Africa
EIA	Environmental Investigating Agency
GHG	greenhouse gases
GDP	Gross Domestic Product
gw	gigawatt
ha	hectares
mt	million tonnes
mw	megawatt
nm	nautical miles
NPP	Nuclear Power Plants

NPT	Nuclear Non-Proliferation Treaty
OFDI	Outward Foreign Direct Investment
Oz	ounces
PPA	Power Purchase Agreements
PRC	People's Republic of China
PGMs	Platinum Group Metals
PPM	Parts Per Million
PV	Photovoltaic
QRD	Quarry Rock Dust
RFE	Russian Far East
REEs	Rare Earth Elements
Sinopec	China Petroleum & Chemical Corp
tw	tetrawatt
USEIA	US Energy Information Administration
WNA	World Nuclear Association

Appendix-1

Countries, Continents, Geographical Regions/Sub-regions

Important changes have taken place in regrouping of the countries following the dissolution of the USSR in 1990s. Some erstwhile confederate states are now part of Europe while others remain in Asia. Falling of Yugoslavia and bifurcation of Czechoslovakia have resulted in carving out of new states; nearly three dozen new countries have been formed since 1990. Here is the updated information:

AFRICA

Eastern Africa

Burundi, Comoros, Djibouti, Eritrea, Ethiopia, Kenya, Madagascar, Malawi, Mauritius, Mayotte, Mozambique, Reunion, Rwanda, Seychelles, Somalia, South Sudan, Uganda, United Republic of Tanzania, Zambia, Zimbabwe.

Middle Africa

Angola, Cameroon, Central African Republic, Chad, Congo, Democratic Republic of the Congo. Equatorial Guinea, Gabon, Sao Tome and Principe.

Northern Africa

Algeria, Egypt, Libya, Morocco, Sudan, Tunisia, Western Sahara.

Southern Africa

Botswana, Lesotho, Namibia, South Africa and Swaziland.

Western Africa

Benin, Burkina Faso, Cabo Verde, Cote d'Ivoire, Gambia, Ghana, Guinea, Guinea-Bissau, Liberia, Mauritania, Mali, Niger, Nigeria, Saint Helena, Senegal, Sierra Leone, Togo.

AMERICAS

Americas is a broader term which could connote Caribbean, Central America, South America and Northern America.

Caribbean

Anguilla, Antigua and Barbuda, Aruba, Bahamas, Barbados, Bonaire, Saint Eustatius and Saba, British Virgin Islands, Cayman Islands, Cuba, Curacao, Dominica, Dominican Republic, Grenada, Guadeloupe, Haiti, Jamaica, Martinique, Montserrat, Puerto Rico, Saint-Barthelme, Saint Kitts and Nevis, Saint Lucia, Saint Martin (French part), Saint Vincent and the Grenadines, Saint Maarten (Dutch part), Trinidad and Tobago, Turks and Caicos Islands, United States Virgin Islands.

Central America

Belize, Costa Rica, El Salvador, Guatemala, Honduras, Mexico, Nicaragua, Panama.

South America

Argentina, Bolivia, Brazil, Chile, Colombia, Ecuador, Falkland Islands (Malvinas), French Guiana, Guyana, Paraguay, Peru, Suriname, Uruguay, Venezuela.

Northern America

Bermuda, Canada, Greenland, Saint Pierre and Miquelon, United States of America.

ASIA

Central Asia

Kazakhstan, Kyrgyzstan, Tajikistan, Turkmenistan, Uzbekistan.

Eastern Asia

China, Hong Kong (Special Administrative Region of China), Macao (Special Administrative Region of China), Democratic People's Republic of Korea, Japan, Mongolia, Republic of Korea.

Southern Asia

Afghanistan, Bangladesh, Bhutan, India, Maldives, Nepal, Pakistan, Sri Lanka.

South-Eastern Asia

Brunei Darussalam, Cambodia, Indonesia, Lao People's Democratic Republic, Malaysia, Myanmar, Philippines, Singapore, Thailand, Timor-Leste.

Western Asia

Armenia, Azerbaijan, Bahrain, Cyprus, Georgia, Iraq, Israel,

Jordan, Kuwait, Lebanon, Oman, Qatar, Saudi Arabia, State of Palestine, Syrian Arab Republic, Turkey, United Arab Emirates, Yemen.

EUROPE

Eastern Europe

Belarus, Bulgaria, Czech Republic, Hungary, Poland, Republic of Moldova, Romania, Russian Federation, Slovakia, Ukraine.

Northern Europe

Åland Islands, Channel Islands, Denmark, Estonia, Faeroe Islands, Finland, Guernsey, Iceland, Ireland, Isle of Man, Jersey, Latvia, Lithuania, Norway, Sark, Svalbard and Jan Mayen Islands, Sweden, United Kingdom of Great Britain and Northern Ireland.

Southern Europe

Albania, Andorra, Bosnia and Herzegovina, Croatia, Gibraltar, Greece, Holy See, Italy, Malta, Montenegro, Portugal, San Marino, Serbia, Slovenia, Spain, the Former Yugoslav Republic of Macedonia.

Western Europe

Austria, Belgium, France, Germany, Liechtenstein, Luxembourg, Monaco, Netherlands, Switzerland.

OCEANIA

Australia and New Zealand

Australia, New Zealand, Norfolk Island.

Melanesia

Fiji, New Caledonia, Papua New Guinea, Solomon Islands. Vanuatu.

Micronesia

Guam, Kiribati, Marshall Islands, Micronesia (Federated States of), Nauru, Northern Mariana Islands, Palau.

Polynesia

American Samoa, Cook Islands, French Polynesia, Niue, Pitcairn, Samoa, Tokelau, Tonga, Tuvalu, Wallis and Futuna Islands.

Other Geographical Connotations

*Sub-Saharan Africa: The term is commonly used to indicate all of Africa except northern Africa. Sudan is also considered in sub-Saharan Africa although it falls in Northern Africa.

*Middle East: A term originated in colonial era still frequently used. The term refers to Arab countries in West Asia and may extend to Turkey, Iran and a few countries in East Africa.

*MENA: The Middle East and North Africa (MENA) is an economically diverse region that includes both the oil-rich economies in the Gulf and countries that are resource-scarce in relation to population, such as Egypt, Morocco, and Yemen. The MENA Region includes: Algeria, Bahrain, Djibouti, Egypt, Iran, Iraq, Israel, Jordan, Kuwait, Lebanon, Libya, Malta, Morocco, Oman, Qatar, Saudi Arabia, Syria, Tunisia, United Arab Emirates, West Bank and Gaza,

Appendix-2

Benchmark Years in Development of Alternative Energy

1861 – Augustin Mouchot invented the first active steam engine which drew energy from the Sun..

1839 – William Grove devised the first hydrogen fuel cell.

1888 – Charles Brush built the first windmill to generate electricity.

1892 – The maiden use of Geo-thermal Energy in Idaho, where geothermal energy was deployed to heat the buildings.

1904 – Larderello Fields of Italy built the first plant to convert geothermal heat into energy.

1908 – Henry Ford built the first mass-produced vehicle, called Ford T Model, running on ethanol. These vehicles remained in production until 1927.

1936 – The world's largest Hoover Dam was built on Colorado River.

1953 – The first solar cell, capable of generating electricity, was made at Bell Laboratories.

1958 – The first solar powered satellite was launched in the USA.

1981-1990 – Over 17,000 wind machines were built in

California.

2006 – The largest offshore wind turbine, called Repower, capable of generating 5 megawatt of electricity, was first installed in the North Sea.

2006 – Geothermal power plant was built in Alaska, which could generate electricity from geothermal energy with the record low temperature of 57°C.

2007 – Ford Fusion Hydrogen 999, powered by hydrogen, reached record speed of over 207 mph.

2009 – The two-seat sun-powered airplane Sky Spark reached the record speed of 250 km/h.